AUSTRALIA AND CYBER-WARFARE

AUSTRALIA AND CYBER-WARFARE

Gary Waters, Desmond Ball and Ian Dudgeon

THE AUSTRALIAN NATIONAL UNIVERSITY

E PRESS

Published by ANU E Press
The Australian National University
Canberra ACT 0200, Australia
Email: anuepress@anu.edu.au
This title is also available online at: http://epress.anu.edu.au/cyber_warfare_citation.html

National Library of Australia
Cataloguing-in-Publication entry
Author: Waters, Gary, 1951-
Title: Australia and cyber-warfare / Gary Waters ; Desmond Ball ; Ian Dudgeon.
ISBN: 9781921313790 (pbk.) 9781921313806 (pdf.)
Series: Canberra papers on strategy and defence ; no. 168
Subjects: Information warfare--Australia.
                   Command and control systems--Australia.
                   Military telecommunication.
Other Authors/Contributors:
                   Ball, Desmond, 1947-
                   Dudgeon, Ian.
                   Australian National University. Strategic and Defence Studies Centre.
Dewey Number: 355.033094

The *Canberra Papers on Strategy and Defence* series is a collection of publications arising
principally from research undertaken at the SDSC. Canberra Papers have been peer reviewed
since 2006. All Canberra Papers are available for sale: visit the SDSC website at <http://rspas.
anu.edu.au/sdsc/canberra_papers.php> for abstracts and prices. Electronic copies (in pdf
format) of most SDSC Working Papers published since 2002 may be downloaded for free from
the SDSC website at <http://rspas.anu.edu.au/sdsc/working_papers.php>. The entire Working
Papers series is also available on a 'print on demand' basis.

Cover design by ANU E Press
Front cover image:
Title: Globe with hand, computer graphic, composition
Artist/photographer: Suk-Heui Park
Image no: 74859163.JPG
Licence holder: Getty Images (royalty-free licence from Matton Images at
<http://www.mattonimages.co.uk/images/jpg/pc_74859163.html>)

Contents

List of Tables

Abstract

This book explores Australia's prospective cyber-warfare requirements and challenges. It describes the current state of planning and thinking within the Australian Defence Force (ADF) with respect to Network Centric Warfare (NCW), and discusses the vulnerabilities that accompany the use by Defence of the National Information Infrastructure (NII), as well as Defence's responsibility for the protection of the NII. It notes the multitude of agencies concerned in various ways with information security, and argues that mechanisms are required to enhance coordination between them. It also argues that Australia has been laggard with respect to the development of offensive cyber-warfare plans and capabilities. Finally, it proposes the establishment of an Australian Cyber-warfare Centre responsible for the planning and conduct of both the defensive and offensive dimensions of cyber-warfare, for developing doctrine and operational concepts, and for identifying new capability requirements. It argues that the matter is urgent in order to ensure that Australia will have the necessary capabilities for conducting technically and strategically sophisticated cyber-warfare activities by the 2020s.

Contributors

Gary Waters spent 33 years in the Royal Australian Air Force, including as a Directing Staff member at the RAAF Staff College; manager of the RAAF's Air Power Studies Centre; and Director Capability Planning in the Australian Defence Headquarters. He has served in London as Head of the Australian Defence Staff; Head of the Theatre Headquarters Project; and Director General Operation *Safe Base* (essentially leading a security operation after the 11 September 2001 terrorist attacks in the United States). Following his retirement (as an Air Commodore) in 2002, Gary Waters became the inaugural Assistant Secretary Knowledge Planning in Defence and, later, Assistant Secretary Information Strategy and Futures within the Chief Information Officer Group, before joining Jacobs Australia as Head of Strategic Initiatives in September 2005. He is currently responsible for its strategic planning and management and reports to the Managing Director. Gary Waters has written 11 books on doctrine, strategy and historical aspects associated with the use of military force, including *Transforming the Australian Defence Force for Information Superiority* (with Desmond Ball) and *Firepower to Win* (with Alan Titheridge and Ross Babbage). Gary Waters is a graduate of a number of institutions: the United Kingdom's Royal Air Force Staff College; the University of New South Wales (MA (Hons) in history); and the Australian Institute of Company Directors. He is a Fellow of the Royal Melbourne Institute of Technology (graduating in accounting and economics) and an Associate of the Australian Society of CPAs. He has recently been awarded his PhD from The Australian National University, where the title of his thesis was 'Networking the ADF for Operations in the Information Age'. A former Fellow of the Australian Institute of Company Directors and Vice President of the United Services Institute, he currently serves as a Board member and Treasurer of the Kokoda Foundation in Canberra.

Desmond Ball is Special Professor at The Australian National University's Strategic and Defence Studies Centre, having been Head of the Centre from 1984 to 1991. Professor Ball is the author or editor of more than 40 books or monographs on technical intelligence subjects, nuclear strategy, Australian defence, and security in the Asia-Pacific region. His publications include *The Boys in Black: The Thahan Phran (Rangers), Thailand's Para-military Border Guards*; *Burma's Military Secrets: Signals Intelligence (SIGINT) from the Second World War to Civil War and Cyber Warfare*; *Signals Intelligence in the Post-Cold War Era: Developments in the Asia-Pacific Region*; *Presumptive Engagement: Australia's Asia-Pacific Security Policy in the 1990s* (with Pauline Kerr); *Breaking the Codes: Australia's KGB Network, 1944–50* (with David Horner); and *Death in Balibo, Lies in Canberra* (with Hamish McDonald). He has also written articles on issues such as the strategic culture in the Asia-Pacific region and defence acquisition programs in the region. Professor Ball was elected a Fellow of the

Academy of Social Sciences of Australia in 1986. He served on the Council of the International Institute for Strategic Studies from 1994 until 2000, and was Co-chair of the Steering Committee of the Council for Security Cooperation in the Asia-Pacific from 2000 until 2002.

Ian Dudgeon is the principal of a Canberra-based consultancy whose services include writing of policy and providing policy advice on and reviews of national security issues. He has served previously in both the Foreign Affairs and Trade and Defence portfolios, and held senior appointments in the Australian Intelligence Community (AIC). This service has included some 12 years on overseas postings in Europe, Asia and the Americas. He is the author of major policy studies for government on Australia's national information infrastructure, AIC support to military operations, and information operations, and of articles published in *Security Challenges*, journal of the Kokoda Foundation, and *Defender*, journal of the Australian Defence Association. He has also lectured at The Australian National University, the Australian Defence Force Academy, University of New South Wales and at other conferences and seminars on national security issues. Ian Dudgeon is presently President of the ACT branch of the Australian Institute of International Affairs, holds a Bachelor of Economics degree from Monash University, is a Graduate of the Australian Institute of Company Directors and has attended a residential Advanced Management Program at Harvard University's Graduate School of Business Administration.

Acronyms and Abbreviations

ACMA	Australian Communications and Media Authority
ACS	Access Control Server
ADDP	Australian Defence Doctrine Publication
ADF	Australian Defence Force
ADGESIM	Air Defence Ground Environment Simulator
ADO	Australian Defence Organisation
ADSCC	Australian Defence Satellite Communications Capability
AEW&C	Airborne Early Warning and Control Aircraft
AFP	Australian Federal Police
AGIMO	Australian Government Information Management Office
AHTCC	Australian High Tech Crime Centre
APEC	Asia-Pacific Economic Cooperation
ASBM	Aerospace Surveillance and Battlespace Management
ASEAN	Association of South East Asian Nations
ASIO	Australian Security Intelligence Organisation
ASIS	Australian Secret Intelligence Service
AusCERT	Australian Computer Emergency Response Team
AWD	Air Warfare Destroyer
BMS–D	Battle Management System–Dismounted
BMS–M	Battle Management System–Mounted
C2	Command and Control
C3I	Command, Control, Communications and Intelligence
C4ISR	Command, Control, Communications, Computers, Intelligence, Surveillance and Reconnaissance
C4ISREW	Command, Control, Communications, Computers, Intelligence, Surveillance, Reconnaissance and Electronic Warfare
CCS	Combat Control System
CERT	Computer Emergency Response Team
CGI	Common Gateway Interface
CIA	Central Intelligence Agency
CIAC	Critical Infrastructure Advisory Council
CII	Critical Information Infrastructure
CIOG	Chief Information Officer Group
CIP	Critical Infrastructure Protection
CIPG	Critical Infrastructure Protection Group
CIS	Combat Information System
CJOPS	Chief of Joint Operations

CNA	Computer Network Attack
CND	Computer Network Defence
CNE	Computer Network Exploitation
CNO	Computer Network Operations
COAG	Council of Australian Governments
COMINT	communications intelligence
COTS	commercial off-the-shelf
CROP	Common Relevant Operating Picture
CRS	Congressional Research Service
CSIAAG	Communications Sector Infrastructure Assurance Advisory Group
DCITA	Department of Communications, Information Technology and the Arts
DDS	Distributed Denial of Service
DFAT	Department of Foreign Affairs and Trade
DGCP	Director General Capability Plans
DGICD	Director General Integrated Capability Development
DHS	Department of Homeland Security
DIGO	Defence Imagery and Geospatial Organisation
DII	Defence Information Infrastructure
DIO	Defence Intelligence Organisation
DISCE	Defence Information Systems Communications Element
DNOC	Defence Network Operations Centre
DPM&C	Department of Prime Minister and Cabinet
DS	Denial of Service
DSD	Defence Signals Directorate
DSTO	Defence Science and Technology Organisation
ECM	electronic countermeasures
ECCM	electronic counter-countermeasures
EDFA	Erbium [Er]-Doped Fibre Amplifier
EIS	Environmental Impact Statement
EOB	electronic order of battle
ESCG	E-Security Coordination Group
ESM	electronic support measures
ESPaC	E-Security Policy and Coordination
EW	electronic warfare
EWOSU	Electronic Warfare Operational Support Unit
FBI	Federal Bureau of Investigation
FJOC	*Future Joint Operating Concept*

GAO	Government Accountability Office
GHz	gigahertz
GII	Global Information Infrastructure
GovCERT.au	Australian Government Computer Emergency Readiness Team
GPS	Global Positioning System
GSMC	Global System for Mobile Communications
HF	high-frequency
HPM	high-power microwave
HQJOC	Headquarters Joint Operations Command
HTTP	Hypertext Transfer Protocol
HUMINT	human intelligence
I&SG	Intelligence and Security Group
IA	Information Assurance
ICT	Information and Communication Technology
IDS	Intrusion Detection Systems
IISS	International Institute for Strategic Studies
IMG	Issue Motivated Group
IMINT	imagery intelligence
infosec	information security
INTERFET	International Force East Timor
IO	Information Operations
IOS	Internetwork Operating System
IP	Internet Protocol
IS&S	Information Superiority and Support
ISO	International Organization for Standardization
ISP	Internet Service Provider
ISR	intelligence, surveillance and reconnaissance
ISS	Internet Security System
IT	Information Technology
ITU	International Telecommunications Union
IW	Information Warfare
IXP	Internet Exchange Point
JDA	Japan Defense Agency
JFCC-NW	Joint Functional Component Command for Network Warfare
JFCC-SGS	Joint Functional Component Command for Space and Global Strike
JIOWC	Joint Information Operations Warfare Center
JORN	*Jindalee* Operational Radar Network
JSF	Joint Strike Fighter

JTF-GNO	Joint Task Force for Global Network Operations
LAN	Local Area Network
malware	malicious software
MASTIS	Maritime Advanced SATCOM Terrestrial Infrastructure System
MDM	Multidimensional Manoeuvre
MHz	megahertz
MND	Minister for National Defense
MRTT	multi-role tanker transport
MTWAN	Maritime Tactical Wide Area Network
NAC	Network Admission Control
NAP	Network Access Point
NAVCAMSAUS	Naval Communications Area Master Station Australia
NAVCOMMSTA	Naval Communications Station Canberra
NCTC	National Counter-Terrorism Committee
NCW	Network Centric Warfare
NCWPO	Network Centric Warfare Project Office
NGO	Non-Governmental Organisation
NIDS	National Illicit Drug Strategy
NII	National Information Infrastructure
NIPRNet	Non-classified Internet Protocol Router Network
NIS	National Intelligence Service
NNWC	Naval Network Warfare Command
NOC	Network Operations Centre
NOSC	National Open Source Center
NSA	National Security Agency
OSINT	Open Source Intelligence
PC	Personal Computer
PDA	Personal Digital Assistant
PIN	Personal Identification Number
PKI	Public Key Infrastructure
PLA	People's Liberation Army
PRC	People's Republic of China
PSIRT	Product Security Incident Response Team
psyop	psychological operation
R&D	research and development
RAAF	Royal Australian Air Force
RAN	Royal Australian Navy
RF	radio frequency

RMA	Revolution in Military Affairs
ROE	Rules of Engagement
RPD	Rapid Prototyping and Development
RPDE	Rapid Prototyping, Development and Evaluation
SASR	Special Air Services Regiment
SATCOM	Satellite communications
Satphone	satellite telephone
SCADA	Supervisory Control and Data Acquisition
SDSS	Standard Defence Supply System
SIGINT	signals intelligence
SIM	Subscriber Identity Module
SIS	Security Impact Statement
SME	Small-to-Medium Enterprise
SMS	Short Message Service
SOCOMD	Special Operations Command
SONET	Synchronous Optical Network
TISN	Trusted Information Sharing Network
UAV	Unmanned Aerial Vehicle
UHF	ultra high-frequency
UNSC	United Nations Security Council
UPS	uninterrupted power systems
URL	Uniform Resource Locator
USSTRATCOM	US Strategic Command
UWB	Ultra-wide band
VHF	very high-frequency
VOIP	Voice Over Internet Protocol
WAP	wireless application protocol
WLAN	wireless local area network

Foreword
by Professor Kim C. Beazley

In 2002 I visited Afghanistan as part of a parliamentary delegation. At Bagram base, while visiting our Special Air Service (SAS) contingent, we were hosted at the headquarters of the 10th Mountain Division on one of the first of their seemingly interminable deployments to the Afghanistan fight. There we saw soldiers sitting behind banks of personal computers controlling everything from the Division's logistics to the units in the field.

We witnessed the interaction between US dominance of the electro-magnetic sphere and its use of cyber-space. Satellites beamed in the ongoing battle and communications relevant to the forces engaged. The Division Commander had the exact location of his forces and those they engaged. We could see the practical effects with orders for A-10 ground support or helicopter extraction as the base responded instantly to requests and constantly added to the information of company and platoon commanders in the field. I asked the Division's Commander how he resisted taking over platoon command in such a situation. "It's difficult," he responded. We were witnessing what the Australian Defence Force described in its 2020 vision statement as 'network-enabled operations'.

It was clear from our subsequent conversations with SAS personnel how much their small unit patrols were enveloped by the plethora of information on their situation which came from a multiplicity of surveillance capabilities and the array of responses and advice they could draw on from the levels of command they were plugged into. It was a privileged view of warriors standing on the bottom rung of future information age warfare.

Politics has distorted what is really important in the Australian debate on our future defence needs. We are obsessed with platforms and personnel numbers. Over the last decade, the Australian Government has burnished its popular security credentials by junking any serious study of platform needs and acquiring capabilities based on immensity with big dollar signs attached, thereby seeking to impress public opinion with size and cost whilst saying little of relevance about modern and future warfighting.

We are a clever and technologically capable people. Partly courtesy of our allied relationship, we are deeply aware of the US ability to exploit the possibilities of electro-magnetic waves and cyber-space. We host and participate in the operation of cutting-edge installations such as those at Pine Gap. Involvement in Iraq (and even more in Afghanistan) and collaboration in the 'war on terror' have given us access to the heart of frontier advances in information operations. Through organisations like the Defence Signals Directorate we make direct contributions. Scattered through Defence and security related departments, like the Attorney-General's, we have institutions responsible for exploiting electro-magnetic and cyber-space for information on those who are our enemies or would be enemies, and using the same space to combat them.

George Orwell said during the Second World War that we sleep safe in our beds because rough men do violence in the night to those who would wish us harm. The rough men are now joined by the 'geeks' of both genders. Yet there are gaps in capability, objectives and missions. As the Australian Government sits down to contemplate its defence White Paper, we expect answers on extra battalions of soldiers, the necessity for the *Super Hornet*, the value of the *Canberra* class LHDs, and the timing of the next generation of submarines. No-one is waiting with bated breath for what it will say about our ability to conduct cyber-warfare or even on what is meant by our capabilities in network-enabled operations. Certainly no-one is waiting to read about our intelligence services transitioning to warfighting operations.

Except perhaps the authors of this study. I cannot begin to attest to the veracity of the material which follows. Like most politicians I have been caught up for a decade in the need for the quick fix in responses to the crises that have emerged since the 11 September 2001 terrorist attacks on the United States.

Whatever emerges in the debate over the next few years on the White Paper, one perception in the White Paper I was responsible for over 20 years ago remains valid today. Our long-term survival depends on a clear understanding of capabilities which may be used against us and on the clear need for a small nation to sustain a technological edge in meeting them. We live in a broader region which is, as one of the authors points out, a test-bed for future information warfare. To this point the edge has been sought defensively; in the future it will need to be sought aggressively.

The authors seek the establishment of an Australian Cyber-warfare Centre to coordinate the development of capabilities that will decisively enhance our forces in the field—but, more than that, ensure that the tools which enhance our warriors are tools in the fight itself. This is a timely book which transcends old debates on priorities for the defence of Australia or forward commitments. It transcends debates about globalism and regionalism. These are global capabilities, but with a multitude of effects in any part of geography that is vital. This book will serve as an invaluable compendium for subsequent judgement about official documents and commentaries that will deluge us as all sections of the Australian Government rethink our national security priorities.

Kim C. Beazley

Professor of Social and Political Theory
University of Western Australia
(Minister for Defence, 1984–90)

March 2008

Chapter 1

Introduction: Australia and Cyber-warfare

Gary Waters and Desmond Ball

In 2005 Air Commodore (Ret'd) Gary Waters and Professor Desmond Ball examined the key issues involved in ensuring that the Australian Defence Force (ADF) could obtain information superiority in future contingencies.[1] The authors discussed force posture, associated command and control systems, information support systems, operational concepts and doctrine. They discussed the ADF's approach to Network Centric Warfare (NCW); examined the command and control aspects of dispersed military operations utilising networked systems; outlined some of the principal strategic, organisational, operational, doctrinal and human resource challenges; and discussed the information architecture requirements for achieving information superiority.

Through its NCW developments, the ADF is aiming to obtain common battlefield awareness and superior command decision-making, using a comprehensive 'information network' linking sensors (for direction), command and control (for flexible, optimised decision-making), and engagement systems (for precision application of force).

The authors examined the twin notions of leveraging off indirect connections and generating effects in unheralded ways to determine what advantages might accrue to the dispersed and networked force and what paradigm shifts would be needed to realise those advantages. They also argued that whatever collaborative form of command and control might be used in future, it had to preserve simplicity, unity of command and balance. This collaborative element would be all about networking, interacting; sharing information, awareness and understanding; and making collaborative decisions.

A number of hypothetical Information Operations (IO) scenarios were presented, in which the ADF was able to defeat an adversary's air assault by cyber-attack; immobilise their naval fleet by electronic warfare (EW) attack; jam and deceive their air defences; destroy or incapacitate their command, control and communications systems; and corrupt their networks. An Information Warfare (IW) architecture for Australia was sketched out and key related issues canvassed.

The authors acknowledged that interrelationships and interdependencies between weapons, sensors, commanders and the supporting network could form the Achilles' heel of the future ADF. Furthermore, as Australia moves down a whole-of-nation approach to security, the way we cooperate and coordinate our activities across government and with allies will extend that supporting network and broaden the potential vulnerabilities.

Reliance on the network will mean enhancing the capability and survivability of Defence's and related networked infrastructures to ensure sustained and protected flows of information. This becomes increasingly problematic as reliance on commercial technologies increases.

Planned US programs offer an unprecedented level of access and availability of information to forces in the field and Australia needs to develop equivalent initiatives so it can 'plug and play' with the United States.

While the authors discussed the Defence Information Infrastructure (DII), which they defined as an 'interconnected, end-to-end set of information systems and technologies that support the electronic creation, collation, processing, protection and dissemination of Defence information', they did not discuss the national and global equivalents, nor how these might be protected. Furthermore, Defence has published several documents since 2005, which underscore its reliance on NCW and the underpinning networks. Hence, it is now timely to examine the potential challenges to those networks and the information that flows across them and how it might all be protected.

Chapter 2 of this volume, by Gary Waters, describes the recent developments with respect to Defence planning for NCW. It examines what the ADF is hoping to achieve through NCW, and outlines the key elements of the *NCW Roadmap* released in 2005 and updated in 2007.

Notwithstanding Defence's published view, the implementation of NCW is being challenged by the demands on planners resulting from the extraordinary tempo of current operations, and the focus on Coalition and regional operations. Much needs to be done in ensuring that Australia will have the necessary capabilities for achieving information superiority around 2020. The work carried out prior to 2005 was fundamentally incomplete as it was mostly concerned with enhancing and sharing battlefield awareness and with shortening decision cycles; it essentially ignored the offensive opportunities and challenges of NCW, and the offensive role of IW more generally. Furthermore, the NCW work since then has continued to pay insufficient attention to the human and organisational dimensions.

The 'war on terror' has stimulated some aspects of IO while further distracting planners from the longer-term construction of an all-embracing NCW architecture that also addresses the offensive and defensive aspects of IW. Recent achievements have been essentially defensive, involving investigative and forensic activities, rather than exploiting cyber-space for offensive IO.

Chapter 3, by Gary Waters, starts by highlighting the value of information to Australia and the ADF today, before discussing the potential forms of IW that could be used against us. There are certain actions an adversary might take against us and certain things we can do to protect ourselves. And there are cyber-crime activities that need to be addressed, as well as critical information infrastructure aspects. This discussion on cyber-attacks and broad network defence sets the scene for the next two chapters on attacking and defending information infrastructures.

Chapter 4, by Ian Dudgeon, discusses how information infrastructures underpin and enable today's information society, and national defence capabilities, and how they shape and influence the way we, and others, live and what we see, think, decide and how we act. It identifies the importance of these infrastructures as targets in war, to achieve physical and psychological outcomes, in order to weaken the military capability and national morale of an adversary, and how psychological outcomes can also strengthen the morale of friends and allies and influence the attitudes of neutral parties. It also discusses how, in certain non-war circumstances, foreign infrastructures may also be targeted to project national power and shape events to national advantage.

Chapter 5, by Gary Waters, discusses the twin challenges of balancing information superiority and operational vulnerability, and security and privacy in information sharing, before examining cyber-security and how we might best secure the Defence and National Information Infrastructures (NIIs). Indeed, this aspect is mentioned in the 2007 *Defence Update*, which stated that: 'There is an emerging need to focus on "cyber–warfare", particularly capabilities to protect national networks to deny information'.[2]

There is a myriad of complex and extremely difficult issues that require resolution before radically new command and control arrangements can be organised, new technical capabilities acquired and dramatically different operational concepts tested and codified. These include the extent to which complete digitisation and networking of the ADF will permit flatter command and control structures; the availability of different sorts of Unmanned Aerial Vehicles (UAVs) and the timeframes for their potential acquisition; the role of offensive operations and the development of doctrine and operational concepts for these; the promulgation of new rules of engagement; and a plethora of human resource issues, including the scope for the creative design and utilisation of reserve forces and other elements of the civil community.[3] These matters will take many years to resolve and even longer, in some cases at least a decade, for the ensuing decisions to be fully implemented.

In chapter 6, Des Ball argues a critical deficiency is the lack of a net-war or cyber-warfare centre. Australia has a plethora of organisations, within and outside Defence, concerned with some aspects of cyber-warfare (including network security), but they are poorly coordinated and are not committed to the full exploitation of cyber-space for either military operations or IW more generally.

A dedicated cyber-warfare centre is fundamental to the planning and conduct of both defensive and offensive IO. It would be responsible for exploring the full possibilities of future cyber-warfare, and developing the doctrine and operational concepts for IO. It would study all viruses, Denial of Service (DS) programs, 'Trojan horses' and 'trap-door' systems, not only for defensive purposes but also to discern offensive applications. It would study the firewalls around computer systems in military high commands and headquarters in the region, in avionics and other weapons systems, and in telecommunications centres, banks and stock exchanges, ready to penetrate a command centre, a flight deck or ship's bridge, a telephone or data exchange node, or a central bank at a moment's notice, and able to insert confounding orders and to manipulate data without the adversary's knowledge. It would identify new capability requirements. It should probably be located in a building close to the Defence Signals Directorate (DSD) in the Russell Hill complex and be run out of the Department of Prime Minister and Cabinet (DPM&C).

ENDNOTES

[1] Gary Waters and Desmond Ball, *Transforming the Australian Defence Force (ADF) for Information Superiority*, Canberra Papers on Strategy and Defence, no. 159, Strategic and Defence Studies Centre, The Australian National University, Canberra, 2005.

[2] See Department of Defence, *Australia's National Security: A Defence Update 2007*, Department of Defence, July 2007, p. 53, available at <http://www.defence.gov.au/ans/2007/pdf/Defence_update.pdf>, accessed 25 February 2008.

[3] Waters and Ball, *Transforming the Australian Defence Force (ADF) for Information Superiority*, pp. 61–68.

Chapter 2

The Australian Defence Force and Network Centric Warfare

Gary Waters

Introduction

The global economy continues to be more networked through information and communication technologies that are fast becoming ubiquitous. Decision-to-action cycles are reducing to cope with the increasing pace of change, which is placing a premium on innovation, information sharing and collaboration. At the same time, national security is being broadened, large quantities of information are flowing along with calls for better quality information, and connectivity is increasing, all of which leads to an increase in the strategic value of information. Ed Waltz expresses it well as:

> the role of electronically collected and managed information at all levels has increased to become a major component of both commerce and warfare. The electronic transmission and processing of information content has expanded both the scope and speed of business and military processes.[1]

In June 2002, Defence released its doctrinal statement on Australia's approach to warfare.[2] In looking at how the Australian Defence Force (ADF) would prepare itself to cope with increasing and rapid change, the focus of the document turned initially to what the Information Age heralded. Attacks on information systems were cited as potential security threats to which the ADF would need to respond.[3] Furthermore, the ADF should expect to find itself increasingly operating in 'small, dispersed combat groups',[4] which would be facilitated in part through technological advances in communications.

Defence also released its long-term vision statement in June 2002—known as *Force 2020*. In articulating a vision of a seamless force—internally with each other (the three Services) and externally with the range of providers, supporting entities and the community[5] —the ADF also highlighted the fundamental need to transform from a platform-centric force to a network-centric one.

The ADF argued that 'the aim of Network-Enabled Operations is to obtain common and enhanced battlespace awareness, and with the application of that awareness, deliver maximum combat effect'.[6] Furthermore, the fundamental building block of networked operations would be a comprehensive 'information network' that linked the sensor grid (for detection), the command and control (C2) grid (offering flexible, optimised decision-making), and the engagement grid (for precision engagement).[7]

Through network-enabled operations, the ADF would be conferred with what it termed 'decision superiority'—'the ability to make better, faster decisions, based upon more complete information than an adversary'.[8] The ADF cited operations in Afghanistan where Unmanned Aerial Vehicles (UAVs) passed real-time targeting information (via video) to aircraft, epitomising the effectiveness of Network Centric Warfare (NCW), through the direct sensor-to-shooter link that allowed rapid engagement of targets. This is what the ADF means by seamless integration of platforms through the information network.[9]

In May 2003, the Chief of the Defence Force, General Peter Cosgrove, noted to an NCW conference that

> while it is likely that some type of crude kinetic effect will still be the ultimate expression of violence in war, it is also likely that as information and network-related war fighting techniques start to mature and to predominate, outcomes will be swifter, as dramatic and paradoxically less bloody than the classic force-on-force attritionist, paradigm of the past.[10]

Indeed, Cosgrove cited the 2003 Iraq War, from which he observed that 'in the main, the Iraqi forces were beaten quickly, spectacularly and comprehensively by a force using what were, on balance, mostly first generation network-centric technologies and concepts'.[11]

The seamless integration called for in *Force 2020* and inferred by Cosgrove has necessitated the ADF moving away from a focus on individual weapon platforms towards exploiting the effectiveness of linked, or networked, forces and capabilities. Networking will allow the sharing of a common and current relevant picture of the operational environment across all components of the joint force. This will, in turn, improve a force's situational awareness, coordination, and importantly, decision-making ability. The joint force will exploit this as it is able to prepare for and conduct operations more smoothly and quickly.

Operations will rely on linking sensors, weapons and commanders, via an appropriate information network, to enable the timely and precise application of military force. By embracing a 'networked' approach to military operations, the ADF will be able to generate greater combat effectiveness than belies its relatively small size—be able to 'punch well above its weight'.

These notions have been reinforced through the release in 2007 of Defence's *Future Joint Operating Concept* (*FJOC*).[12] The *FJOC* starts with Air Chief Marshal Angus Houston's vision for the ADF, which is to be 'a balanced, networked and deployable force, staffed by dedicated and professional people, that operates within a culture of adaptability and excels at joint, interagency and coalition operations'.[13] The *FJOC* goes on to argue that the force must operate in the seamless manner described in *Force 2020*, not only to maximise the ADF's collective warfighting capabilities but also its ability to operate with interagency and coalition partners. Improved networking will enhance the ADF's capability advantage over potential adversaries as it also relies on its people to generate the underlying capability advantage and the 'knowledge edge' needed for the future.

Increasingly, the ADF must be capable of both executing effective combat operations and providing military support to national responses in more complex environments. The ADF must move to develop a hardened, networked, deployable joint force that is characterised by adaptability and agility to handle the full range of military operations across the full spectrum of conflicts.

In the information dimension, future adversaries will utilise informal communications technologies that are cheap, ubiquitous and difficult to trace, and increasingly secure and sophisticated networked C2 and intelligence, surveillance and reconnaissance (ISR) systems, leveraging commercial satellite capabilities and improved geospatial information.

The *FJOC* adopts a national effects-based approach, which involves taking a whole-of-nation view of security to find the most appropriate tool to achieve national objectives—the military is but one of the tools. It is underpinned by the NCW Concept that will help link ADF, Australian and coalition sensors, engagement systems and decision-makers into an effective and responsive whole. NCW seeks to provide the future force with the ability to generate tempo, precision and combat power through shared situational awareness, clear procedures, and the information connectivity needed to synchronise friendly actions to meet the commander's intent.

The ADF and Defence will work in cooperation with other government and non-government agencies (where appropriate) to develop the capability for an integrated multi-agency response capability, extending the network to other agencies as appropriate.

In the Future Warfighting Concept,[14] the ADF adopted Multidimensional Manoeuvre (MDM) as its approach to future warfare. MDM seeks to negate the adversary's strategy through the intelligent and creative application of an effects-based approach against an adversary's critical vulnerabilities. It uses an indirect approach to defeat the adversary's will, seeking to apply tailored strategic responses to achieve the desired effects.

MDM operations are designed to focus on specific and achievable effects through integrating joint warfighting functions (force application, force deployment, force protection, force generation and sustainment, C2, and knowledge dominance). A fundamental of MDM is the ability to employ NCW and operate in joint task force, interagency and/or coalition arrangements to conduct effective operations. The joint operational concept underlying MDM are best described in terms of the ability to *reach*, *know* and *exploit* as follows:

- *Reach*—Reach describes the future force's ability to operate in multiple dimensions both inside and outside the operational area and across the physical, virtual and human domains in order to understand and shape the environment; deter, defeat and deny the adversary; and provide military assistance in support of national interests. Reach is best accomplished as part of an integrated whole-of-government approach across the spectrum of military, diplomatic, economic and informational actions.[15]
- *Know*—The future force will build and sustain sufficient knowledge from national and international sources to allow it to identify required actions and assess the effects of those actions. It will understand itself and its capabilities, those of its adversaries, as well as the operating environment, which will enable the force to better carry out those actions that create decisive effects. Information is at the base of knowledge dominance, and knowing requires that the future force is able to utilise and integrate information from strategic, operational and tactical sources, both military and civilian. However, information must be turned into knowledge that is timely, relevant and accurate. This knowledge must be acquired, prioritised, refined and shared across the strategic, operational and tactical levels and within the joint force and as part of multi-agency and multinational efforts.[16]
- *Exploit*—The future force will integrate its joint capabilities with other elements of national power in order to achieve effects in support of national strategic objectives. Effects are the outcomes of the actions taken to change unacceptable conditions and behaviours, or to create freedom of action to achieve desired objectives. The force will identify, create and exploit effects through acquiring knowledge and establishing reach. To exploit its capability to produce effects, the future force will continually assess its effects and adjust its actions to take into account the iterative interaction of military, diplomatic, economic and informational actions that are taken as part of Australia's whole-of-government approach.[17]

The following attributes define the ADF of the future:[18]

- *Balanced*—The future force must possess an appropriate mix of capabilities in order to mount the range of operations envisaged. It must offer a multiplicity of responses and not rely on 'niche' capability.

- *Networked*—The future force will need assured access to other agency, coalition and open-source information. The ability to operate effectively will be contingent on the integrated forces' networks and decision-making infrastructures, early warning systems, communications, environmental monitoring and positional data. Adversaries may exploit any vulnerability in the nation's network to undermine cohesion and effectiveness.

- *Deployable*—In the future, the ADF will need to operate at a distance from established bases in Australia, either independently or with coalition forces, potentially involving deployments with regional or global reach. Force elements will need to be configured and prepared for short-notice deployments that can be sustained with limited infrastructure support. This will require either a capability to lift forces into the contingency area or basing rights close to the contingency area. A forced-entry capability will also be critical to the ADF's ability to respond.

- *Integrated and Interoperable*—The ADF must continue the transition to a force (with fully integrated services) that is interoperable with other agencies of the government and its coalition partners and allies. Legacy systems should, to the extent possible, be made to function in the integrated environment until replaced. As the degree of integration and synchronisation is increased, new training and systems will need to be established. Military capabilities should be designed to be interoperable from conception, not as an afterthought in the development process.

- *Survivable and Robust*—Each element of the future force must be able to protect itself against the range of existing and evolving threats. Timely investment in lower signatures, protection, countermeasures and redundancy to match likely threats will be required.

- *Ready and Responsive*—The future ADF must observe, anticipate and be prepared to serve Australia's global interest in an evolving strategic and geopolitical situation.

- *Agile and Versatile*—The future ADF must be able to respond rapidly to a diverse range of missions and tasks. This will require versatile forces that are tailored and scalable for deployment. They will need an ability, the extent of which will be dictated by force structure, to re-form, reconstitute, regroup and re-engage, especially during periods of concurrent operations.

- *Precise and Discriminating*—The goal for future operations is to achieve precise effects, with minimum planning and response time, from a distance if required. For the future ADF, precision must not be limited to the application of kinetic force, but also be incorporated into executing information operations (IOs) and minimising unintended consequences. While traditional technology will initially provide the potential to improve precision, emergent technology must be used to support widespread cross-platform responses that ensure maximum flexibility and discrimination.

Enhanced discrimination capabilities will permit high-value targets to be struck with greater certainty.

- *Lethal and Non-lethal*—The ADF must increase its capability to produce desired effects through the considered and coordinated use of both lethal and non-lethal methods, using both kinetic and non-kinetic means. These effects will be enhanced by leveraging technology advances which improve precision and discrimination, and by employing a whole-of-nation approach.

- *Persistent and Poised*—Persistence ensures that the joint force has the required endurance at all levels to generate and deploy forces for long periods, while poise ensures that critical fighting elements are within range of a potential target area. Persistence incorporates force protection, logistics, infrastructure development and sustaining the capacity of ADF people to work and fight. The persistence of the future ADF may necessitate a greater level of force dispersal, leading to a requirement to generate effects from dispersed locations, while at the same time being poised to project force at short notice. Poise is achieved through either expanding deployability or securing basing rights close to likely contingency areas.

- *Sustainable*—The increasing mobility, tempo and changeability of future force operations will require an adaptive, modular, network-enabled logistic system operating in a contiguous and non-contiguous mission space.

- *Capable of Concurrency*—The future force must be able to conduct operations in more than one location simultaneously. The Defence Planning Guidance provides guidance on the number and nature of deployed operations across the maritime, land, air and space environments. The major capabilities underpinning these operations will be the effective use of information to C2 forces, the ability to conduct strike operations, and the ability to generate and sustain military forces.

- *Legal and Ethical*—In accordance with ADF core values, the ADF operates within the Australian legal framework and the international Law of Armed Conflict. The future ADF must continue to take pride in operating within an ethical framework, derived from a strong warfighting tradition.

Armed with those insights into where the ADF is headed, this chapter discusses the ADF's NCW Concept, *NCW Roadmap*, and Information Superiority and Support (IS&S) Concept in more detail to set the ensuing discussion in subsequent chapters on the 'cyber' dimension, Information Warfare (IW), how information infrastructures can be targeted, how they can be protected, and how both offensive and defensive IOs can best be brought together via an Australian cyber-warfare centre.

The ADF'S NCW Concept

In December 2003, the final NCW Concept Paper was produced by the Policy Guidance and Analysis Division, within the Strategy Group.[19] The Concept

Paper argued that NCW involved the linkage of engagement systems to sensors through networks and the sharing of information between force elements. Information is only useful if it allows people to act more effectively: this makes the human dimension fundamental to NCW. NCW thus has two closely related and mutually reinforcing dimensions—the human dimension and the network dimension. The NCW Concept argues that

> the human dimension is based on professional mastery and mission command, and requires high standards of training, education, doctrine, organisation and leadership. This dimension is about the way people collaborate to share their awareness of the situation, so that they can fight more effectively. It requires trust between warfighters across different levels, and trust between warfighters and their supporting agencies.[20]

The Concept continues:

> The second dimension, the network, connects major military systems, including engagement, sensor and command systems. The network dimension was the initial focus of development, but change here was always expected to have a profound influence on the human dimension.[21]

NCW is seen by the ADF as a 'means to realising a more effective warfighting ability. New technology will change the character of conflict, but war's enduring nature—its friction, fog and chaotic features—will persist'.[22] The Australian NCW Concept accepts this enduring nature of war, but does seek to reduce the effects of fog and friction.

The purpose of the NCW Concept was to provide a starting point for the identification and exploitation of the opportunities of NCW. It would inform and shape the conduct of Defence's NCW-related research and experimentation programs, which would further crystalise an understanding of the opportunities and risks associated with NCW. The ADF would continually revisit the concept in order to confirm its validity based on the lessons learned through research, experimentation and operational experience.

The ADF's NCW Concept is based on the following premises, which will be tested through experimentation:[23]

- Professional mastery is essential to NCW.
- Mission command will remain an effective command philosophy into the future.
- Information and intelligence will be shared if a network is built by connecting engagement systems, sensor systems, and C2 systems.
- Robust networks will allow the ADF, and supporting agencies, to collaborate more effectively and achieve shared situational awareness.

- Shared situational awareness will enable self-synchronisation, which helps warfighters to adapt to changing circumstances and allows them to apply MDM more effectively.

The last two are fundamental in transforming the way in which information is managed, used and exploited. These are expanded on below.

Networks

Robust networks involve sharing information and intelligence through a connected network that also includes engagement, sensor and command systems. While we might look at these systems separately, many of the ADF's platforms perform across all four grids. There is an expectation that NCW will offer an ability to explore alternatives, whereby sensors may be separated from the engagement system, or the ADF might be able to reduce the size of its deployed force.

The ADF aims to develop and integrate an advanced sensor system, ranging from space-based assets to humans, to gather widely disparate information. In doing this, the ADF expects a certain amount of redundancy (without wasteful duplication) to ensure persistent battlespace awareness. That said, the integration of information from sensors will not provide complete understanding of the battlespace, although greater analytic capacity to produce intelligence is anticipated. In the end, commanders will still have to decide whether to fight for more information or to work with the information and intelligence available.[24]

Advanced command support systems will bring together information about the adversary, own and friendly forces, other parties, and the environment into a Common Relevant Operating Picture (CROP). In addition, these systems will allow different levels of the ADF, relevant government agencies, and coalition partners to work together. Through these advanced command support systems, the ADF would expect to enhance its capacity for mission rehearsal, wargaming and development, and analysis of possible courses of action. Essential logistic information between the warfighters and support bases should also be able to be exchanged more effectively.[25]

In terms of its engagement systems, the ADF will aim for its decision-makers to have timely access to the most useful engagement systems for the mission, noting that different systems have different levels of mobility, firepower and self-protection. The intent will be to shorten the time between detection, identification, engagement and assessment. The network dimension of NCW assists forces to:[26]

- *collect* relevant information;
- *connect* units and platforms through networking, doctrine, training and organisation;

- *use* the information and intelligence in a timely manner to achieve the commander's intent; and
- *protect* the network from external interference or technical failure.

The *collect-connect-use-protect* framework is the means through which the ADF can organise its effort to develop the network. This framework, which underpins the IS&S Concept, is discussed in more detail later.

The ADF will need to monitor carefully the way networks are progressing in the commercial sector, where developments will have a strong influence on what is available, noting that Defence will move increasingly to commercial off-the-shelf (COTS) solutions for its hardware and software.

As networks and people come together and the notions of trust and information sharing become integral to making decisions, the ADF will need to be aware of interactions across the information, cognitive and physical domains:[27]

- In the *information domain*, connectivity allows people to share, access and protect information.
- In the *cognitive domain*, connectivity allows people to develop a shared understanding of the commander's intent, and to identify opportunities in the situation and vulnerabilities in the adversary.
- In the *physical domain*, selected elements of a force are equipped to achieve secure and seamless connectivity and interoperability. This connectivity will allow some sensor systems to pass target acquisition information directly to engagement systems. Based on the shared understanding developed in the other domains, forces are able to synchronise actions in the physical domain.

These domains also apply to an adversary; hence, the NCW Concept also seeks to influence an adversary by disrupting their ability to function effectively within, and across, each of these domains.

Shared situational awareness

Shared situational awareness develops as people absorb information, collaborate to understand its implications, and then acquire a shared view of the situation at hand. Thus, shared situational awareness brings together both the network and human dimensions of NCW.[28]

Collaboration is essential to shared situational awareness because it allows widely dispersed forces to use their battlespace awareness for mutual advantage in terms of analysis, decision-making, and application of force. The challenge for the ADF will be to cope with a shift from sequential planning activities through a hierarchy to an ongoing interaction between different levels, which will save time and provide opportunities for simultaneous action. Again, both the network (technical means) and human dimensions (ability of people) are important for collaboration.[29]

Collaboration requires a high degree of trust throughout the chain of command. Hence, training and personnel development must provide opportunities for different elements of the ADF to become familiar with one another, the Defence organisation more broadly, and with other agencies.

Self-synchronisation

Another challenge of NCW will be for the ADF to evolve from its top-down way of synchronising forces and actions. People will need to use their shared situational awareness to recognise changes and opportunities themselves, and to act without direction to meet the commander's intent. Self-synchronisation will thus lead to speedier decision-to-action cycles by capitalising on the shared understanding and collective initiative of lower-level commanders and staffs.[30]

Balancing risks and opportunities

NCW will focus on warfighting through the concept of MCM. The network is only an enabler to warfighting effectiveness; it supplements but cannot replace the skill, intuition and willpower of the ADF's people. The focus on training, doctrine, leadership and organisation will balance the technical aspects that often dominate discussion of NCW. The Concept identified five areas of potential risk:[31]

- The failure to incorporate the human dimension into thinking about NCW.
- The potential for disruption—through an adversary exploiting vulnerabilities, indirect attacks on networks, denial of communications, or misleading information. Network integrity will need to be assured.
- Pursuit of a 'transparent' battlespace, which is almost certainly unachievable. The ADF must not expect NCW to deliver an 'unblinking eye' across the whole battlespace. Commanders must cope with, and thrive in, ambiguity.
- The potential exists to be overloaded with information, threatening friendly forces with self-induced paralysis. Commanders may also become addicted to information, causing hesitation while waiting for the key piece of evidence.
- Commanders could attempt to micro-manage operations.

The real opportunities presented by NCW offer priorities and benchmarks for further development. These include people, operations, logistics, decision-making, training, organisation, doctrine, and major systems as follows:

- NCW will help the ADF's *people* conduct their individual and collective tasks better.

- NCW will help to make a small force like the ADF more efficient and effective on *operations*. NCW should assist the ADF to operate in a more dispersed manner, while permitting the concentration of combat power when required.
- NCW will allow technology to be used to automate *logistic* reporting, support sophisticated self-diagnostic systems that improve equipment reliability, and improve service delivery in areas such as medical support.
- NCW will help ADF commanders to make better *decisions* by improving their ability to command operations, control forces and conduct planning.
- The improved ability to conduct *training and education* will help to increase the confidence and skills of individuals, and enhance trust between individuals, even when they work in dispersed organisations.
- The ADF will use NCW to improve the *organisation*'s ability to shape or react to evolving situations, to collaborate better across organisational boundaries, and to exploit collective knowledge.
- The ADF will need to respond to an NCW-induced fast rate of change by adopting trial *doctrine*, which could leverage off lessons learned from experimentation, training and operations.
- Adopting a network-centric approach is intended to reduce incompatibilities between and within *major systems*, and allow each to be employed with maximum effect. New systems will need to fit seamlessly within an information infrastructure. Legacy systems that remain will need to be adapted for NCW. This has implications for coalition operations and for cooperation with other government agencies; hence clear standards will be crucial.[32]

The NCW Concept clearly supports *Force 2020* by teasing out these implications of NCW on the Fundamental Inputs to Capability. *Force 2020* argued:

> Our strategic advantage will come from combining technology with people, operational concepts, organisation, training and doctrine. We must be careful to ensure that technology does not give an illusion of progress—we cannot afford to maintain outdated ways of thinking, organising and fighting.[33]

FJOC continues in this theme, adding the ability to reach, know and exploit (as discussed earlier).

The *NCW Roadmap*

The 2020 networked force will be an exceptionally complex organisation with a range of different relationships (machine and human) which will require careful and thorough integration. Conducting rapid prototyping and development (RPD) activities will allow Defence to mitigate the risks that are inherent in this integration. It will support experimentation and provide the ability to simulate future capabilities to aid in determining the optimum level of integration between engagement, sensor and C2 systems. RPD will reduce the risk of implementing NCW and help accelerate change.[34]

The capacity to concurrently 'learn by doing'[35] is also important for implementing NCW. Hence, the way ahead, or the *NCW Roadmap*, centres on a 'learn by doing' strategy. A draft *NCW Roadmap* was also produced in December 2003, to plan and coordinate the implementation of NCW, but was not formally released until October 2005. This document set out the future requirement for NCW, the current level of networked capability, and the steps needed to realise a future networked force.[36] Defence also released a short document explaining NCW.[37]

The first step in 'learn by doing' involves constructing the foundation for enhanced collaboration and shared situational awareness. This foundation will, in the main, comprise the information infrastructure and the governance measures needed to improve connectivity for selected force elements. As this underlying infrastructure improves; so too will the collaborative ability of elements within the ADF.

As stated earlier, the long-term aspiration is to link all ADF elements into a 'single virtual network', where information is assembled and passed through a series of interlinked grids: sensor grids gather data; information grids fuse and process it; and engagement grids (overlaid by an appropriate C2 grid) allow warfighters to generate the desired battlespace effects.

NCW has the potential to facilitate the collaboration required for the ADF to employ MDM more effectively. It will assist the force to generate the tempo, agility and ultimately the warfighting advantage needed to prevail against a wide variety of adversaries. However, the ADF will also need to enhance the capacities of its people and its platforms for the future, ensuring they are networked to better exploit the chaotic conditions of the battlespace. The fog, friction and ambiguity will remain; the ADF must ensure that it is better able to exploit this than an adversary.[38]

The 2005 *NCW Roadmap* provides the direction, and initial steps, to implement the NCW Concept. It is Defence's guide to discovering and exploiting the opportunities of NCW; and identifies four key actions:[39]

- set the NCW-related targets for Defence to achieve;
- establish the Network to provide the underlying information infrastructure upon which the networked force will be developed;
- explore the human dimensions of the networked force and initiate changes in doctrine, education and training with appropriate support mechanisms; and
- accelerate the process of change and innovation through the establishment of a RPD capability in partnership with Industry.

Subsequent to publication of the 2005 *NCW Roadmap*, Defence released a publication entitled *Explaining NCW*. [40] Defence argues in this document that 'NCW is a means of organising the force by using modern information technology (IT) to link sensors, decision-makers and weapon systems to help people work more effectively together to achieve the commander's intent'.[41] Furthermore, NCW can 'contribute significantly to producing a warfighting advantage'.[42] The information network sits at the centre, linking the C2 systems, sensor systems, and engagement systems.

Defence identified the key capability development projects that would deliver the desired network capability and packaged them as a system, which also provided a model for future systems planning and capability integration. This included projects associated with communications in the maritime, land and air environments; a wide area communications network; network management and defence; satellite communications (SATCOM); tactical information exchange; information exchange in a coalition environment; and other information protection measures.[43]

Indeed, General Peter Cosgrove had already presaged some of these projects when he said, in May 2003,[44] that Defence would look at how it could harmonise sophisticated technology with people in networked systems, noting that the future maritime surveillance and response project and the Joint Command Support System showed great promise here. He also referred to other behind-the-scenes changes, such as the adoption of a standardised 'J series' message format (supporting tactical information exchange) as being critical pieces of the NCW puzzle.

He also foreshadowed that Defence would place key links, such as Airborne Early Warning and Control (AEW&C) aircraft, into the network over the next few years. And the recent move into space, through the Optus satellite, provides another part of an increasingly pervasive network. These systems, said Cosgrove, will magnify the pay-off from our network-centric approach.[45]

A governance system for the Defence Information Environment was implemented in 2003, led by the newly-created Office of the Chief Information Officer, to ensure that activity across the Department was aligned and enforced. This governance framework encapsulated the interrelationships and interdependencies in the development, management and operation of Defence's supporting information environment.[46]

Network protection was to be designed through the use of an Information Security Architecture, which is an integral component of a Defence-wide Information Architecture. The architecture would address the design of information systems, access control, data management and accountability frameworks.

The human dimension

The examination of NCW's human dimension was to focus on the following areas:

- the nature of C2;
- transitioning to a new way of operating;
- nurturing innovation; and
- Defence culture.

As General Cosgrove had remarked in May 2003, 'it is vital that we keep people in our focus as we implement NCW. Anybody can buy technology. Anybody can copy concepts. But nobody can duplicate the advantage we get from our smart, dedicated and adaptable people'.[47]

The 2005 *NCW Roadmap* developed this human dimension further, with a focus on doctrine, education, training and development. It argued the need to raise NCW awareness, educate the senior leaders, prepare the future leaders, understand the future workforce, produce the NCW tools and plan for doctrine development, and develop a mechanism for evaluation and feedback of lessons learnt.[48]

Accelerating change and innovation

In terms of accelerating change and innovation, Defence was to establish a RPD program by July 2004 to fulfill three functions:

- identify and test new technologies, concepts, procedures and organisations that could be implemented in the near term (6–18 months) to improve the ADF's networked warfighting capabilities;
- identify early problems with the implementation of NCW and use RPD as an intervention activity to redress the problem or mitigate risk; and
- provide for the rapid delivery of capability to warfighters to meet or anticipate emerging security challenges.

A Rapid Prototyping, Development and Evaluation (RPDE) capability was set up in 2005, with the mission of enhancing 'ADF warfighting capacity through accelerated capability change in the NCW environment'.[49] Importantly, RPDE allows collaboration across a wide range of organisations, more rapid fielding of capability improvements, and a focus on all fundamental inputs to capability—personnel, organisation, collective training, major systems, supplies, facilities, support, and command and management.[50]

Defence's Information Superiority and Support Concept

In August 2004 Defence released its IS&S Concept, which articulated the key components of the concept, described the architectural approach to be taken, identified the target states for the future and posed a key set of questions to be addressed.[51]

The key components of the concept (focused on connecting, collecting, using and protecting information) are outlined below:[52]

- *Ubiquitous network or information distribution (Connect)* enables effective sharing of information by people, systems, applications and sensors, whether it is by voice, data or video. It involves coordinating the infrastructure including fixed and mobile communications, computers, processes, systems and tools that enable the sharing of information throughout the force to achieve information and decision superiority. When the network is threatened, it must allow for a graceful degradation of service availability and access to information that ensures continuity of operations. Once the threat has been resolved, access to information and original service availability levels must be rapidly reconstituted.
- *Persistent awareness (Collect)* enhances situational awareness that allows better perception of battlespace elements in terms of time and space, the comprehension of their meaning and projected intent. Persistent awareness tools and systems help collect, collate and fuse disparate data and information, which requires greater attention to content management. Persistence does not mean continuous; it means sufficient awareness to enable the ADF to act in a way that is operationally responsive and appropriate.
- *Smart use or decision support (Use)* focuses on achievable intent. It involves planning and the provision of information and common processes and collaboration tools to facilitate timely and effective decision-making throughout the various levels of command.
- *Pervasive Security (Protect)* provides a secure information environment that offers a trusted and reliable flow of information to continuously support operations and business activities.[53]

Networking issues

The IS&S Concept emerged from the observation that networking improves efficiency and effectiveness of operations. It depends on computers and communications to link people through information flows, which in turn depends on interoperability across all systems. Networking involves collaboration and sharing of information to ensure that all appropriate assets can be quickly brought to bear by commanders during combat operations.[54] In a 2004 report by the US Congress, a number of key networking issues were identified, the most relevant of which, for the purposes of this discussion, was network architectures.[55]

Because NCW is so highly dependent on the interoperability of communications equipment, data, and software to enable the networking of people, sensors, and platforms (both manned and unmanned), network architectures are very important. Architectures are needed to bring together all the elements of NCW technology that rely on line-of-sight radio transmission for microwave or infra-red signals, or laser beams; as well as other technologies that aggregate information for transmission through larger network trunks for global distribution via fibre-optic cables, microwave towers, or both low-altitude and high-altitude satellites.[56]

The ADF architecture must enable rapid communications between individuals in all three Services, as well as rapid sharing of data and information between mobile platforms and sensors used across the ADF. As the ADF comes to depend on networking, the network itself must be able to re-form when any communications node is interrupted (the United States refers to this as dynamically self-healing).

The ADF's capability planning for NCW

The ADF's concept is less about warfighting and more about how net-centric capability will enable future warfighting.[57] Defence has been moving steadily along the net-centric path for several years now in terms of developing the capability to provide the ability for data to be exchanged across linked networks. Some ships and aircraft as well as fixed and deployable communications systems have already achieved a degree of data connectivity.[58]

The ADF can deliver secure C2 to small-scale deployments around the world. It can use its 'Secret' and 'Restricted' fixed networks, as well as data-links (Link 11) and radios (*Parakeet*) to provide certain levels of connectivity.

It can draw information from multiple sources such as the over-the-horizon *Jindalee* Operational Radar Network (JORN) and radars on Royal Australian Navy (RAN) ships to generate a basic level of situational awareness. Major combat units are linked by voice communications, with some aircraft and ships data-linked. Secure SATCOM are available to select elements of the ADF, such as Special Forces, and secure satellite data-link provides connectivity between ground-based air defence systems and the Royal Australian Air Force (RAAF)'s Regional Operations Centre.[59] Defence has argued that it is developing its networked force by:[60]

- creating new doctrine, better training and a more agile organisation, all of which leads to people operating more effectively as a network;
- guiding force development through the release of the 2005 *NCW Roadmap* and Integration Plan (the latter is not available publicly);
- connecting broad areas of Defence so that information can be shared and used more cooperatively; and
- fast-tracking the introduction of new technology through its RPDE program.

The Chief of Capability Development Group is responsible for implementing NCW across Defence, which reinforces the earlier point that the ADF's network-centric focus is on developing capability to enhance future warfighting effectiveness.

The major network projects that have been identified for implementation thus far include:[61]

- A joint command support environment, which will link the air, maritime, land and special operations elements into the one single ADF command system.
- A similar project, which will integrate intelligence systems.
- Military SATCOM that will provide the ADF with coverage throughout the region.
- Tactical information exchange that will facilitate movement of information from sensors to weapon systems, starting with *ANZAC* frigates and *Hornet* fighters.
- Battlespace communications for air, land and maritime forces, which will provide the information backbone and tactical data distribution for deployed forces.
- A Defence wide-area communications network, which will provide the next-generation fixed infrastructure for secure computers and telephones. Linkages will be established between the fixed and deployable communications networks.
- A Defence network operations centre, which will enhance current computer network defence capabilities.
- Combined information exchanges with the United States, United Kingdom, Canada and New Zealand, which will provide a permanent system for exchanging information and enabling collaboration.
- A number of major systems that will connect to the 'network', such as the New Air Combat Capability, high-altitude long-endurance UAVs, the Air Warfare Destroyer (AWD), and a suite of other projects that will 'harden' and 'network' the Australian Army.

By 2020, a networked ADF should be able to generate a range of lethal and non-lethal effects that are timely, appropriate and synchronised. It will have continuous information connectivity to link fighting units, sensors and decision-makers that sees an ADF with increased situational awareness and the capacity to act decisively. The Defence C2 system will promote collaboration. Defence will be capable of rapidly deploying and protecting an optimised force. A pervasive network of active and passive sensors will improve awareness for force protection purposes. Key logistics networks will be linked and offer connectivity and collaboration.[62]

Defence has adopted a systems approach to improve the integration of many complex projects. Capability Development Group is focusing on three key areas of development in its Capability Plan. These are:

- the enabling infrastructure to deliver the robust communications network;
- the enabling information systems to support mission command, ISR, imagery and military geospatial information sharing; and
- the combat platforms and hardware to deliver combat effects enabled by the information systems and infrastructure.[63]

There has been some significant slippage in the milestones needed to evolve NCW capability that were articulated in the 2005 *NCW Roadmap*. The key milestones in that Roadmap were as follows:[64]

- 2008: Broadband Networked Maritime Task Group—initial capability.
- 2008: Networked Aerospace Surveillance and Battlespace Management (ASBM) capability.
- 2009: Interim Networked Land Combat Force.
- 2010: Networked Fleet—mature capability.
- 2010: Integrated Coalition Network capability.
- 2012: First Networked Brigade.
- 2013: Networked Air Warfare Force.
- 2014: Second networked Brigade.
- 2015: Robust Battlespace Network.
- 2015: Networked Joint Task Force.

The milestones in the 2007 *NCW Roadmap* are as follows:[65]

- 2008: Networked Special Operations Unit.
- 2008: Networked Air Combat Force.
- 2009: Networked Battle Group.
- 2009: Networked Rapid Mobility Force.
- 2011: Networked Maritime Task Group.
- 2011: Networked Combat Support Force.
- 2011: Networked Tactical ISR.
- 2012: First Networked Brigade.
- 2012: Networked Special Operations Task Group.
- 2012: Networked Deployable Joint Task Force Headquarters.
- 2014: Networked Fleet.
- 2014: Second Networked Brigade.
- 2014: Networked Aerospace Command, Control, Communications, Computers, Intelligence, Surveillance, Reconnaissance and Electronic Warfare (C4ISREW) Force.
- 2014: Networked Operational ISR.
- 2014: Networked Deployable Joint Task Force.

- 2014: Networked Coalition Combat Force.
- 2016: Networked Joint Force.

Notwithstanding the slippages, there has been some acceleration, and the 2007 *NCW Roadmap* certainly provides both a clearer and a more comprehensive view. For a start, the new Roadmap breaks these milestones into six domains as depicted in the following table:[66]

Table 1—NCW Domain and Milestone Overview

Coalition	Joint Force	Maritime	Land	Aerospace	ISR
2014: Networked Coalition Combat Force	2012: Networked Deployable Joint Task Force Headquarters	2011: Networked Maritime Task Group	2008: Networked Special Operations Unit	2008: Networked Air Combat Force	2011: Networked Tactical ISR
	2014: Networked Deployable Joint Task Force	2014: Networked Fleet	2009: Networked Battle Group	2009: Networked Rapid Mobility Force	2014: Networked Operational ISR
	2016: Networked Joint Force		2012: First Networked Brigade	2011: Networked Combat Support Force	
			2012: Networked Special Operations Task Group	2014: Networked Aerospace C4ISREW Force	
			2014: Second Networked Brigade		

(*Source*: Director General Capability and Plans, *NCW Roadmap 2007*, Defence Publishing Service, Canberra, March 2007, p. 22.)

A second improvement in the 2007 *NCW Roadmap* is that it breaks down each of the six domains into the grids—C2 capability, network capability, sensor capability, and engagement capability. Furthermore, in the two years between the roadmaps, Defence had come to understand the need for greater cooperation and coordination across projects to deliver a new networked ADF for 2016. These milestones are discussed in more detail below under each domain.

Maritime

- *Networked Maritime Task Group—2011*: This capability will be principally delivered through three projects: JP 2008 Phase 3F (Military SATCOM Capability); SEA 1442 (Maritime Tactical Wide Area Network); and AIR 5276 (the AP-3C *Orion* upgrade). The outcomes of these projects—Maritime Advanced SATCOM Terrestrial Infrastructure System (MASTIS) terminals, wide area Local Area Networks (LANs), and new high-speed data-links for the AP-3C *Orion*—will provide broadband connectivity for major fleet units.[67]

- *Networked Fleet—2014*: SEA 1442 (Maritime Communications and Information Management Architecture Modernisation) equipment acquisition will enable Internet Protocol (IP) networking at sea between major fleet units via the expansion of the Maritime Tactical Wide Area Network (MTWAN). Phase 4 of SEA 1442 will deliver upgraded communications capabilities through replacement radios, antennas and other systems and will form the basis of the networked fleet. SEA 4000 (AWD) and JP 2048 (Helicopter Landing Dock) will significantly improve the warfighting capability of this force.[68]

Land

- *Networked Special Operations Unit—2008*: JP 2097 Phase 1A (Project *Redfin*) will deliver a Special Operations Vehicle and Command, Control, Communications, Computers, Intelligence, Surveillance and Reconnaissance (C4ISR) package to provide a network-enabled capability for Special Air Services Regiment (SASR) land operations.[69]
- *Networked Battle Group—2009*: This force will consist of a number of all arms sub-units based on a mechanised infantry battle group headquarters or a cavalry battle group headquarters. The force will be equipped with digital communications and battle management systems.[70]
- *First Networked Brigade—2012*: The Networked Battle Group 2009 capabilities will be extended to complete the rollout to other Brigade elements. This milestone will see the evolution of the 1st Brigade into a fully networked capability. The major activities during this period will include the introduction of a Battle Management System–Dismounted (BMS–D) (Land 125), introduction of the M1A1 (Land 907), new communications bearers (JP 2072 Phase 2 and 3), Battle Management System–Mounted (BMS–M) (Land 75 Phase 4) and improved logistics support (JP 2077).[71]
- *Networked Special Operations Task Group—2012*: Project *Redfin* (JP 2097 Phase 1B) and the delivery of JP 2030 (Special Forces Command Support Capability) will provide communications, sensors, C2 and engagement systems for a complete Special Operations Task Group. This will consist of Special Operations Command (SOCOMD) units and external units in direct support. Gateway interfaces will enable the exchange of information with other ADF and coalition networks.[72]
- *Second Networked Brigade—2014*: The Networked Brigade 2012 capabilities will be extended to a second Brigade (3rd Brigade). This milestone will result from the capability delivery of two major projects, LAND 75 (Battlefield Command Support System) and LAND 125 (Solider Combat System).[73]

Aerospace

- *Networked Air Combat Force—2008*: The provision of Tactical Data Links to the *ANZAC* frigate and F/A-18 *Hornet* (through JP 2089 Phase 2) combined with the force multiplier capabilities of the AIR 5077 (AEW&C aircraft) are key enablers in networking the Aerospace Domain.[74]

- *Networked Rapid Mobility Force—2009*: The delivery AIR 8000 Phase 3 (four Boeing C-17 *Globemaster III*) combined with AIR 5402 (Multi-Role Tanker Transporters—Airbus A330-200) will provide significantly increased capacity and range for operations. The new Mobile Regional Operations Centre (AIR 5405) will provide an enhanced deployable C2 capability that can be combined with the transportable Tactical Air Defence Radar System (AIR 5375) to improve the networking and protection of the rapid mobility assets.[75]

- *Networked Combat Support Force—2011*: The delivery of further applications for the JP 2030 Phase 8 (Joint Command Support System) coupled with advances to the Standard Defence Supply System (SDSS) provided by JP 2077 (Improved Logistics Information Systems) will enhance the ability to provide timely levels of resources to deployed forces.[76]

- *Networked Aerospace C4ISREW Force—2014*: Projects JP 5077 (AEW&C), AIR 7000 Phase 1B (Multi-mission Unmanned Aerial System), and the upgraded AP-3C *Orion* (AIR 5276) will provide highly capable long-range ISR sensors. The delivery of the JP 2065 Phase 2 (Integrated Broadcast System) will disseminate ADF and allied ISR and Blue Force Tracking information to deployed forces via Tactical Data Links (JP 2089) and SATCOM (JP 2008). The delivery of SEA 4000 (AWD) will bring with it for the first time Theatre Ballistic Defence capabilities as well as an improved ability to control airspace well beyond Australian landmass where required.[77]

ISR

- *Networked Tactical ISR—2011*: Project DEF 7013 Phase 4 (Joint Intelligence Support System) provides intelligence to commanders from networked databases and applications. The Battlespace Communications System (Land) (JP 2072) and Maritime Communications and Information Management Architecture Modernisation (SEA 1442 Phase 4) projects provide the means for tactical information dissemination. Project AIR 7000 Phase 1B (Multi-role UAV) and JP 129 (Tactical UAV) will provide new tactical and strategic sensors.[78]

- *Networked Operational ISR—2014*: By 2014, common geospatial information data and products will be accessible to users across the network (JP 2064 Phase 3). Advanced operational ISR capabilities will provide enhanced views of the battlespace through the space-based surveillance capability delivered by JP 2044 Phase 3 and the JORN upgrade (JP 2025 Phase 5). The ability to conduct tactical electronic warfare (EW) will be improved through Force Level Electronic Warfare DEF 224 (*Bunyip*). The networking of the ISR capabilities, together with the means to fuse the information, will be delivered by JP 2096. The Multi-mission Maritime Patrol Aircraft and the Multi-mission

Unmanned Aerial System (AIR 7000 Phase 1B and Phase 2B) will complete the acquisition of advanced networked collection systems from the tactical to the strategic levels. The networking of Defence into national and coalition intelligence and ISR networks will be accomplished by the remaining deliverables of JP 2096.[79]

Joint force

• *Networked Deployable Joint Task Force Headquarters—2012*: For Joint Operations Command, Projects JP 8001, JP 2008 and JP 2030 Phase 8 will enhance situational awareness and connectivity. Other enhancements include JP 2089 (Tactical Data Links) and JP 2065 (Integrated Broadcast System). The delivery of the Mobile Regional Operations Centre (JP 5405) will enhance the C2 capability of deployable headquarters.[80]

• *Networked Deployable Joint Task Force—2014*: Improved spaced-based surveillance (JP 2044 Phases 3A and 3B), various communications bearers (JP 2008, SEA 1442, JP 2072), and surveillance projects (JP 2025, AIR 5432, AIR 7000 Phase 1B) will contribute to enhanced joint task force operations. The delivery of the Amphibious Ships (JP 2048 Phase 4A/4B) will further enhance the C2 capability of deployable headquarters.[81]

• *Networked Joint Force—2016*: By 2016 the ADF will have achieved the infrastructure, tools and C2 systems (JP 2030, AIR 5333, and JP 2072) capable of providing a robust battlespace network across the whole ADF. Communications beyond line-of-sight will be improved through JP 2008 Phase 4. The achievement of this milestone will allow the ADF to conduct NCW operations, thereby greatly improving warfighting flexibility and combat effectiveness.[82]

Coalition

• *Networked Coalition Combat Force—2014*: This milestone coincides with the achievement of one key milestone from each of the other five Domains:[83]
  • Networked Fleet 2014;
  • Second Networked Brigade 2014;
  • Networked Aerospace C4ISREW Force 2014;
  • Networked Operational Intelligence 2014, and
  • Networked Joint Task Force 2014.

The Networked Coalition Domain focuses on how Defence integrates with Australia's allies and other government agencies. The integrated Coalition Network Capability will allow for the seamless integration of ADF C2, ASBM and communications into established coalition network architectures. A mature Networked Coalition Domain will also provide significant whole-of-government benefits through integrating Defence with a range of other government agencies.[84]

Conclusion

Defence is responding to the challenges of networking the future force in several ways, one of which is through an integrated Defence Information Environment. This is an environment where the Defence Information Infrastructure (DII) will need to be increasingly developed and managed as an integrated entity. Data, user applications, common information services, user devices, systems hardware, networks and data-links, and communications bearers will be integrated to form a foundation or backbone information capability.

A second and equally important way is through Capability Development Group's integrated capability development approach, which includes an NCW Program Office. Here, all capability projects are looked at through the lens of the particular domain (maritime, land, aerospace, ISR, joint force, and coalition); and then through the lens of the type of capability (C2, network, sensor, or engagement).

The degree of integration will be based on developing a robust communications network, supported by consistent joint doctrine and comprehensive training that should address and remove operational, intelligence and single Service stove-pipes. The approach is, of necessity, an incremental one that needs to be synchronised over time and defined through comprehensive architectures and technical standards, all underpinned by a focus on high levels of interoperability—particularly with Australia's close partners (especially the United States).

To achieve this, Defence will need to renew its efforts to better coordinate all aspects of NCW—the capability projects, the DII, the human element, the organisational element, the role of industry through RPDE, and research and development (R&D)—as well as further develop its NCW compliance process.

Delivery of the following would mean that a networked deployable joint task force should be possible by 2014, provided there is no substantial program slippage:

- key capability projects for maritime communications;
- military SATCOM;
- satellite surveillance;
- Mobile Regional Operations Centres;
- the ADF Air Defence System (known as *Vigilare*);
- the Joint Battlespace Communications System;
- the Battlefield Command Support System;
- Maritime Communications;

- the Tactical Information Exchange Domain (data-links);
- the Joint Command Support System;
- Integrated Broadcast System;
- Joint Intelligence Support System; and
- Project *Redfin*;

as well as the major capability projects such as:

- the Navantia-designed F100 AWD;
- the UAVs;
- the F/A-18 *Hornet* upgrade, the F/A-18E/F *Super Hornet*;
- the Headquarters Joint Operations Command (HQJOC) Project;
- the A330 multi-role tanker transport (MRTT) aerial refueling tankers;
- the *Wedgetail* AEW&C aircraft;
- the F-35 Joint Strike Fighter (JSF);
- the M1A1 *Abrams* tank;
- the soldier combat system program;
- the large *Canberra* class amphibious ships; and
- a network maintenance and upgrade program.

The ADF has learned a lot from recent operational experiences and experimentation, and is applying that to good effect.

In the networked ADF of the future, based on the capabilities outlined above, transparency of information and self-synchronisation must become key characteristics. This means that the various cultures and sub-cultures in Defence will have to converge, language will have to standardise, and collaboration will have to be the norm.[85]

NCW is as much an organisational and workforce phenomenon as it is a technological one. The ADF needs to prepare both its people and its organisation for the transition to this new technological base.

The omens are strong for achievement of a networked ADF in 2016, and realisation of the NCW vision. However much remains to be done, and serious intellectual effort needs to be devoted to realise an effectively networked ADF of the future. The capability milestones must be adhered to, the Network and its underlying infrastructure must be established, the human dimension must be developed (together with new doctrine, training and education programs), new organisational structures and processes must evolve, and the process of change and innovation must be accelerated through increased use of industry, experimentation and a more coherent and focused research program.

ENDNOTES

[1] Edward Waltz, *Information Warfare: Principles and Operations*, Artech House Publications, Boston and London, 1998, p. 2.

[2] Department of Defence, *The Australian Approach to Warfare*, Department of Defence, Canberra, June 2002, available at <http://www.defence.gov.au/ publications/taatw.pdf>, accessed 4 March 2008.

[3] Department of Defence, *The Australian Approach to Warfare*, p. 12.

[4] Department of Defence, *The Australian Approach to Warfare*, p. 26.

[5] Department of Defence, *Force 2020*, Department of Defence, Canberra, June 2002, p. 17, available at <http://www.defence.gov.au/publications/f2020.pdf>, accessed 25 February 2008.

[6] Department of Defence, *Force 2020*, p. 19.

[7] Department of Defence, *Force 2020*, p. 19.

[8] Department of Defence, *Force 2020*, p. 20.

[9] Department of Defence, *Force 2020*, p. 20.

[10] Speech by General Peter Cosgrove to the Network Centric Warfare Conference on 20 May 2003, entitled 'Innovation, People, Partnerships: Continuous Modernisation in the ADF', available at <http://www.defence.gov.au/cdf/speeches/past/speech20030520.htm>, accessed 25 February 2008.

[11] General Peter Cosgrove, speech entitled 'Innovation, People, Partnerships: Continuous Modernisation in the ADF', available at <http://www.defence.gov.au/cdf/speeches/past/speech20030520.htm>, accessed 25 February 2008.

[12] Released as Department of Defence, *Joint Operations for the 21st Century*, Department of Defence, Canberra, May 2007, available at <http://www.defence.gov.au/publications/FJOC.pdf>, accessed 25 February 2008.

[13] Department of Defence, *Joint Operations for the 21st Century*, p. iii.

[14] Department of Defence, *Future Warfighting Concept*, Australian Defence Doctrine Publication (ADDP)-D.02, Department of Defence, Canberra, 2003, available at <http://www.defence.gov.au/publications/fwc.pdf>, accessed 25 February 2008.

[15] Department of Defence, *Joint Operations for the 21st Century*, p. 15.

[16] Department of Defence, *Joint Operations for the 21st Century*, pp. 15–16.

[17] Department of Defence, *Joint Operations for the 21st Century*, p. 16.

[18] Department of Defence, *Joint Operations for the 21st Century*, pp. 20–22.

[19] This work, that involved close collaboration across Defence was led by the then Head of the Policy Guidance and Analysis Division, Air Vice-Marshal John Blackburn. It was published as *Enabling Future Warfighting: Network Centric Warfare* (ADDP-D.3.1) in February 2004, but not released until May 2004.

[20] Department of Defence, *Enabling Future Warfighting: Network Centric Warfare*, ADDP-D.3.1, Australian Defence Headquarters, Canberra, February 2004, p. v.

[21] Department of Defence, *Enabling Future Warfighting: Network Centric Warfare*, p. v.

[22] Department of Defence, *Enabling Future Warfighting: Network Centric Warfare*, p. v.

[23] Department of Defence, *Enabling Future Warfighting: Network Centric Warfare*, p. 2-2.

[24] Department of Defence, *Enabling Future Warfighting: Network Centric Warfare*, p. 2-4.

[25] Department of Defence, *Enabling Future Warfighting: Network Centric Warfare*, p. 2-4 to p. 2-5.

[26] Department of Defence, *Enabling Future Warfighting: Network Centric Warfare*, p. 2-5.

[27] Department of Defence, *Enabling Future Warfighting: Network Centric Warfare*, p. 2-6.

[28] This is a fundamental point and I am indebted to Lieutenant Colonel Mick Ryan for his insight and persistence in teasing this out.

[29] Department of Defence, *Enabling Future Warfighting: Network Centric Warfare*, p. 2-6.

[30] Department of Defence, *Enabling Future Warfighting: Network Centric Warfare*, p. 2-7 for further elaboration.

[31] Department of Defence, *Enabling Future Warfighting: Network Centric Warfare*, p. 3-3 for expansion.

[32] Department of Defence, *Enabling Future Warfighting: Network Centric Warfare*, p. 3-4 to p. 3-6.

[33] Department of Defence, *Force 2020*, p. 11.

[34] Department of Defence, *Enabling Future Warfighting: Network Centric Warfare*, p. 4-1.

[35] Department of Defence, *Enabling Future Warfighting: Network Centric Warfare*, p. 4-2 for greater elaboration on this notion of 'learn by doing'.

[36] Department of Defence, *NCW Roadmap*, Department of Defence, Canberra, October 2005; updated in Director General Capability and Plans, *NCW Roadmap 2007*, Defence Publishing Service, Canberra, March 2007, available at <http://www.defence.gov.au/capability/ncwi/docs/2007NCW_Roadmap.pdf>, accessed 25 February 2008.

[37] Department of Defence, *Explaining NCW*, Department of Defence, Canberra, 21 February 2006, available at <http://www.defence.gov.au/capability/NCWI/docs/Explaining_NCW-21feb06.pdf>, accessed 25 February 2008.

[38] Department of Defence, *Enabling Future Warfighting: Network Centric Warfare*, p. 5-1.

[39] See Director General Capability and Plans, *NCW Roadmap 2007*, p. v.

[40] Department of Defence, *Explaining NCW*.

[41] Department of Defence, *Explaining NCW*, p. 5.

[42] Department of Defence, *Explaining NCW*, p. 5.

[43] The NCW milestones are described in Director General Capability and Plans, *NCW Roadmap 2007*, pp. 22–31.

[44] General Peter Cosgrove, 'Innovation, People, Partnerships: Continuous Modernisation in the ADF'.

[45] General Peter Cosgrove, 'Innovation, People, Partnerships: Continuous Modernisation in the ADF'.

[46] Patrick Hannan, as the first Defence Chief Information Officer, showed great insight and leadership in bringing an architectural approach and strong governance to the Defence Information Environment, ably supported by his chief architect, John Sheridan, and Sheridan's team of experts.

[47] General Peter Cosgrove, 'Innovation, People, Partnerships: Continuous Modernisation in the ADF'.

[48] Director General Capability and Plans, *NCW Roadmap 2007*, pp. 14–15.

[49] Department of Defence, *NCW Roadmap*, p. 32.

[50] Department of Defence, *NCW Roadmap*, p. 34. RPDE is discussed in some detail in NCW Roadmap 2007, pp. 43–45.

[51] Department of Defence, *A Concept for Enabling Information Superiority and Support*, Department of Defence, Canberra, August 2004.

[52] These core elements have been agreed by Defence and released publicly in *A Concept for Enabling Information Superiority and Support*, p. 11.

[53] Colleagues Andrew Balmaks, Mike Banham, Jason Scholz, and Anna McCarthy worked with me in developing these definitions, prior to public released of the IS&S Concept.

[54] US Department of Defense, *Report on Network Centric Warfare*, 2001, available at <http://www.defenselink.mil/nii/NCW/ncw_sense.pdf>, accessed 25 February 2008; and Ret. Admiral Arthur Cebrowski, Speech to Network Centric Warfare 2003 Conference, January 2003, available at <http://www.oft.osd.mil>, accessed 25 February 2008.

[55] Congressional Research Service (CRS) Report for Congress (received through the CRS Web), entitled 'Network Centric Warfare: Background and Oversight Issues for Congress', 2 June 2004, by Clay Wilson (Specialist in Technology and National Security Foreign Affairs, Defense, and Trade Division), available at <http://www.fas.org/man/crs/RL32411.pdf>, accessed 25 February 2008.

[56] CRS Report for Congress, 'Network Centric Warfare: Background and Oversight Issues for Congress'.

[57] See Department of Defence, *Explaining NCW*, p. 9.

[58] Department of Defence, *Explaining NCW*, p. 15.

[59] Department of Defence, *Explaining NCW*, pp. 15–16.

[60] Department of Defence, *Explaining NCW*, p. 17.

[61] Department of Defence, *Explaining NCW*, Figure 2.2, p. 19.

[62] Department of Defence, *Explaining NCW*, p. 18. These target states for 2020 relate to the future warfighting functions described in Department of Defence, *Future Warfighting Concept*, pp. 36–38.

[63] Director General Capability and Plans, *NCW Roadmap 2007*, p. 21.

[64] Department of Defence, *NCW Roadmap*, p. 20.

[65] Director General Capability and Plans, *NCW Roadmap 2007*, p. 22.

[66] Director General Capability and Plans, *NCW Roadmap 2007*, p. 22.

[67] Director General Capability and Plans, *NCW Roadmap 2007*, pp. 23–24.

[68] Director General Capability and Plans, *NCW Roadmap 2007*, p. 24.

[69] Director General Capability and Plans, *NCW Roadmap 2007*, p. 25.

[70] Director General Capability and Plans, *NCW Roadmap 2007*, p. 25.

[71] Director General Capability and Plans, *NCW Roadmap 2007*, pp. 25–26.

[72] Director General Capability and Plans, *NCW Roadmap 2007*, p. 26.

[73] Director General Capability and Plans, *NCW Roadmap 2007*, p. 26.

[74] Director General Capability and Plans, *NCW Roadmap 2007*, p. 27.

[75] Director General Capability and Plans, *NCW Roadmap 2007*, p. 27.

[76] Director General Capability and Plans, *NCW Roadmap 2007*, p. 27.

[77] Director General Capability and Plans, *NCW Roadmap 2007*, p. 27.

[78] Director General Capability and Plans, *NCW Roadmap 2007*, p. 28.

[79] Director General Capability and Plans, *NCW Roadmap 2007*, p. 28.

[80] Director General Capability and Plans, *NCW Roadmap 2007*, pp. 29–30.

[81] Director General Capability and Plans, *NCW Roadmap 2007*, p. 30.

[82] Director General Capability and Plans, *NCW Roadmap 2007*, p. 30.

[83] Director General Capability and Plans, *NCW Roadmap 2007*, p. 31.

[84] Director General Capability and Plans, *NCW Roadmap 2007*, p. 31.

[85] David Schmidtchen, *The Rise of the Strategic Private: Technology, Control and Change in a Network Enabled Military*, The General Sir Brudenell White Series, Land Warfare Studies Centre, Canberra, 2006, p. 49, where he refers to the four cultures as Navy, Army, Air Force, and Australian Public Service, and the sub-cultures as trades and professions.

Chapter 3

Information Warfare—Attack and Defence

Gary Waters

Introduction

Information is used to create value and achieve a desired end-state or effect. Preventing this value from being realised, on the one hand, and protecting those systems that allow that value to be realised, on the other, are caught up in the notion of Information Warfare (IW). This chapter addresses these two aspects—the value of information and IW. It discusses the methods an adversary might use to attack Australia's networks and other capabilities and what we should do to prevent that. Cyber-crime is the other side of the same coin—posing a threat to our networks. We need to determine just what constitutes our Critical Information Infrastructure (CII) in Australia and ensure we have adequate protection measures in place. These aspects are also canvassed.

The value of information

The objective for using information in business is to create capital value; across government, it is to deliver value to the public; and the objective in the military is to achieve a desired end-state or effect. Thus, information is used to create value and achieve a desired end-state or effect. ·

In a business sense, the value of Information Technology (IT) can be exploited by gaining leverage through process innovation, by applying data and information in one process to other processes, and by sharing networks or selling excess capacity.[1] This also applies to the fixed information infrastructure that supports the Australian Defence Organisation (ADO). Ed Waltz extends this thinking to the military operations dimension by arguing that leverage can be gained through the use of data-links to deliver real-time targeting information to weapons; intelligence could be applied to support the competitiveness of the economy; and coalition networks with appropriate security mechanisms can burden share.[2]

The real utility of information can be seen as a function of its accessibility and flexibility, as well as its reliability, together with the way in which it contributes to achieving a desired end-state or effect.

In traditional military thinking, information has tended to be viewed as battlefield intelligence and tactical attacks on enemy radar and telephone networks. That thinking should now be broadened to view information as a powerful lever that can alter an adversary's high-level decisions. Indeed, it becomes a strategic asset, in which opposing sides will try to shape the other's actions by manipulating the flow of intelligence and information.[3]

It was out of this thinking that *Command and Control Warfare* emerged in the early 1990s. David Ronfeldt and John Arquilla of RAND Corporation in Santa Monica took this further into the realms of cyber-war—turning the balance of information and knowledge in one's favour.[4]

In a strategic sense, the value of information boils down to the ability to 'acquire, process, distribute and protect information, while selectively denying or distributing it to adversaries and/or allies'.[5] In other words, the real value comes from providing the right information to the right people at the right time in the right place and in the right form.

While information or knowledge superiority might win wars, it is also highly fragile—as Alvin and Heidi Toffler say, 'a small bit of the right information can provide an immense strategic or tactical advantage. The denial of a small bit of information can have catastrophic effects'.[6] This leads to the notion of *information superiority*, which is the capability to collect, process, and disseminate an uninterrupted flow of information while exploiting or denying an adversary's ability to do the same.[7]

Should an information attack be launched against the Australian Defence Force (ADF), it should be able to take defensive or offensive measures. A defensive response would generate alerts, increase the level of access restrictions, terminate vulnerable processes, or initiate other activities to mitigate potential damage on the ADF. An offensive response would support targeting and specific attack options that the ADF might wish to carry out.[8]

With many weapons increasingly coming to rely on information—such as smart munitions that use Global Positioning System (GPS) guidance—the ADF can expect information to become more directly relevant in warfare of the future. Similarly, a digitised force should be able to operate at a higher tempo than a non-digitised one through its improved ability to coordinate actions.[9]

Open source information

Outside the traditional military realm, the explosive growth of personal computers (PCs) and their linking via the Internet offers vast quantities of public information that is freely available. While the intelligence community is probably the most affected, all branches of government are impacted. Indeed, the ongoing extremist threat, other non-State threats, and the increasing need for whole-of-government consideration of security issues should be sharpening the government's focus on the potential and indeed the strategic and tactical importance of open source information on the Internet.

Joseph Nye, a former head of the US National Intelligence Council in the 1990s stated:

> Open source intelligence is the outer pieces of the jigsaw puzzle, without which one can neither begin nor complete the puzzle ... open source intelligence is the critical foundation for the all-source intelligence product, but it cannot ever replace the totality of the all-source effort.[10]

Within this context, the Australian Government needs to consider the value of open source information; the importance of the ever-increasing amount of information on the Internet; the utility of new analytic tools that can collect, sift, analyse, and disseminate this publicly available information; and, training issues relating to open source technology and techniques.[11]

Open source information may be defined as that information which is publicly available and that anyone can lawfully obtain by request, purchase, or observation. However, the acquisition of such information must conform to any existing legal copyright requirements. Open source information can include:

- media such as newspapers, magazines, radio, television, and computer-based information;
- public data such as government reports, and official data such as budgets and demographics, hearings, legislative debates, press conferences, and speeches;
- information derived from professional and academic sources such as conferences, symposia, professional associations, academic papers, dissertations and theses, and experts;[12]
- commercial data such as commercial imagery; and
- grey literature such as trip reports, working papers, discussion papers, unofficial government documents, proceedings, preprints, research reports, studies, and market surveys.[13]

It can also include company proprietary, financially sensitive, legally protected, or personally damaging information that is unclassified.[14] Increasingly, it also encompasses information derived from Internet blogs.

In 2004, the US Congress called for an open source centre that could collect, analyse, produce, and disseminate open-source intelligence. Congress argued that open source intelligence was a valuable source of information that had to be integrated into the intelligence cycle to ensure that US policy-makers were fully informed. Accordingly, the National Open Source Center (NOSC) was established on 1 November 2005, and placed under the management of the

Central Intelligence Agency (CIA). The NOSC's functions included 'collection, analysis and research, training, and IT management to facilitate government-wide access and use'. The intent now is to provide a centre of expertise for exploiting open-source information across whole-of-government. Indeed, the NOSC can be tasked by other agencies for specific research.[15]

For Australia, it will be important to ensure that open-source experts are available across all government agencies so as to also avoid unnecessary duplicative efforts. Thus, some form of Centre with the requisite expertise and capability to train open source experts in government agencies is needed now. Such a Centre could improve information sharing across agencies by using state-of-the-art IT, seeking to maximise connectivity throughout the Australian Government and eliminate incompatible formats and any duplicative effort. Individual agencies should still be able to maintain independent open-source databases, but they would have to be maintained in formats accessible to other agencies.

Commercial satellites offer a good supplement to imagery from government satellites. Indeed, today, anyone with access to the Internet can obtain high-quality overhead imagery. It would be important for an Australian Centre, therefore, to have links to the Defence Imagery and Geospatial Organisation (DIGO).

In short, there would seem to be real merit in establishing an Australian Open Source Agency outside the Intelligence Community, with the intent to provide open-source information to all elements of the Australian Government, including parliamentary committees. Increasingly, open-source information will become essential for all functions of government and will demand more concerted efforts to acquire and analyse the vast quantities of available information. This could be one of the functions of a Cyber-warfare Centre as described in the final chapter by Des Ball.

Information Warfare

In the end, information is an important enabler, which may at times be of great strategic value, but in essence this is usually because of other actions, effects, and end-states to which it contributes.

Some of the pro-information literature tends to argue that information dominance avoids the need to use force and that it leads to an ability to disrupt the adversary rather than destroy his forces. While there may be an element of truth in that, the use of force is not incompatible with achieving a superior information position, and disruption and destruction are not mutually exclusive.[16]

War will continue to be a dangerous and violent clash, while improved information will tend to facilitate a more economical use of force.[17] Information is not an end in itself.[18] Rather, it is a means to an end, and increasingly nations will view that end as the achievement of an effect, whether it be diplomatic, military, economic, informational, societal, technological, or a combination of these instruments of national power.

Ed Waltz offers a basic model of warfare in terms of options for attack. He argues that one can launch a physical attack, engage in deception, carry out a psychological attack, or engage in an information attack.[19] In each of these, information has a key role. The aim of these options is to destroy, to deceive or surprise, to disorient, and to severely dislocate (by affecting confidence in information through destruction, deception or disorientation).

The logical extension of all of this for the country that can master the information domain is to close the loop on Sun Tzu's observation that the acme of skill is to 'subdue the enemy without fighting'.[20] The US thought leader Dick Szafranski updated Sun Tzu's observation by arguing that the knowledge systems of an adversary should be the primary strategic target.[21] This is not meant to imply that information is the only option for attack, but rather that its importance as a partner to the more traditional physical forms of attack has increased.

In a warfighting sense, sensor technologies have extended the engagement envelope; computers and communications technologies have led to an increase in the tempo of operations; and the integration of sensors into weapons has made them more precise and lethal. The real transformation, therefore, has not been in sensor, weapons or IT *per se*, but in shifting the focus from the physical dimension to the information dimension.

Waltz calls this the transition toward the dominant use of information and the targeting of information itself. He makes a neat distinction between IW, which emphasises the use of information as a weapon or target, and Information-based Warfare, which he describes as the use and exploitation of information for advantage—often in support of physical weapons and targets.[22]

Martin Libicki contends that there are seven different types of IW that can be categorised by the nature of the operations they contain. These are:[23]

- *Command and Control Warfare*, which is aimed at separating command from the forces by attacking command and control (C2) systems.
- *Intelligence-based Warfare*, which aims to support other forms of attack by collecting, exploiting and protecting information.
- *Electronic Warfare*, which attacks communications by concentrating on transfer (radio-electronics) and formats (cryptographic).
- *Psychological Warfare*, which attacks the human mind.
- *Hacker Warfare*, which ranges over the Global Information Infrastructure (GII).

- *Economic Information Warfare*, which aims to control an economy by controlling certain information.
- *Cyber Warfare*, which brings together abstract forms of terrorism, simulation and reality control.

Waltz differentiates *Command and Control Warfare* as attacks against the Defence Information Infrastructure (DII) and *Information Warfare* as attacks against the National Information Infrastructure (NII).[24]

The key aspects that emerge from Waltz's analysis of this work are that three security-related attributes are needed from an information infrastructure—availability, integrity and confidentiality.[25] Availability encompasses information services (processes) as well as information itself (context). An adversary's objective in IW would be to disrupt (availability), corrupt (integrity) or exploit (confidentiality or privacy).

The ADF is developing new ways of accomplishing its missions by leveraging the power of information and applying network-centric concepts, made possible by rapidly advancing IT. Indeed, military leaders have always recognised the key contribution that information makes to victory in warfare. To that end, they have always sought to gain a decisive information advantage over their adversaries.

As David Alberts argues, in the Information Age militaries now need to understand

> the complex relationships among information quality, knowledge, awareness, the degree to which information is shared, shared awareness, the nature of collaboration, and its effect on synchronisation, and turning this understanding into deployed military capability.[26]

Technological advances in recent years have vastly increased the military's capability to collect, record, store, process, disseminate, and utilise information. However, the advances in the ability to process information simply have not kept pace with the ability to collect that information.

Technology is also bridging distances and providing the capability for individuals to be able to interact with each other in increasingly sophisticated ways, making it easier for individuals and organisations to share information, to collaborate on tasks, and to synchronise actions.

Thus, our increasing reliance on satellite technology and IT to mount joint and combined operations presents any adversary with an opportunity for an asymmetric advantage if our networks can be corrupted, damaged, or destroyed. Furthermore, our national reliance on IT networks and the increasing interconnectedness of all forms of national power mean that a strategic campaign could be waged against both our military forces deployed in distant theatres and our domestic IT infrastructure.

How would an adversary attack us?

As a Center for Security Policy paper observed recently: 'The increasing digitization of military operations, economic and financial infrastructure, as well as all modern communication networks carries with it a great risk'.[27] The combination of global connectivity, mobility of employees, and rapid technological change exposes Australia's civilian information infrastructure to certain risks such as fraud, theft, industrial espionage and disruption to business continuity. Thus, the civilian IT systems of our nation and the ADF's Command, Control, Communications, Computers, Intelligence, Surveillance and Reconnaissance (C4ISR) systems are at great risk if they are not adequately defended.

What would a state-based adversary seek to do though? Such an adversary would seek to disrupt our decision-making process by interfering with our ability to obtain, process, transmit, and use information.[28] Furthermore, depending on the circumstances, that adversary may wish to impede our decision-making with the aim of delaying or even deterring conflict so that it is no longer on our terms. Once the use of military force becomes likely, the adversary would use IW to shape the battlespace in such a way that their likelihood of victory would be improved.

The adversary would support the training and operations of their military with civilian computer expertise and equipment, sourced from training and education establishments and IT industries in their country. The intention would be to integrate these civilians into regular military operations. Then these integrated IW capabilities would be directed primarily at the ADF.

One avenue of attack could be through computer viruses designed to attack our computer systems and networks. A virus such as Myfip could be used as it can be disguised and, once activated on poorly protected network systems, can compromise the entire network information system and steal any of the following file types:[29]

- .pdf—Adobe Portable Document Format
- .doc—Microsoft Word Document
- .dwg—AutoCAD drawing
- .sch—CirCAD schematic
- .pcb—CirCAD circuit board layout
- .dwt—AutoCAD template
- .dwf—AutoCAD drawing
- .max—ORCAD layout
- .mdb—Microsoft database.

The network could lose its documents, plans, communications and databases, or it might simply have all of its proprietary information stolen and remain totally unaware.

The adversary would test our cyber-defences with low-level assaults and incursions, essentially conducting 'cyber-reconnaissance' by probing the computer networks of Australian Government agencies and private companies, seeking to identify weak points in the networks, understand how Australian leaders think, discover the communications patterns of Australian Government and private companies, and obtain valuable information stored throughout the networks.[30]

An attack could be launched against ADF or other Government Department email systems through a Distributed Denial of Service (DDS), where systems would be overwhelmed by 'botnets' that make a request for service from a single information resource. These botnets (that could number more than 100 000)[31] would be extensive networks of computers used by the adversary to overload the response capability of our information systems.

The adversary will have been planning for several years to conduct a limited war by attacking our C4ISR systems, as well as our national economic system, possibly using a terrorist organisation to detonate an electro-magnetic pulse weapon above Australia, at the appropriate time. The adversary would also have developed its IW doctrine along the following lines:

> Information Warfare involves combat operations in a high-tech battlefield environment in which both sides use information-technology means, equipment, or systems in a rivalry over the power to obtain, control and use information. Information Warfare is combat aimed at seizing the battlefield initiative; with digitized units as its essential combat force; the seizure, control, and use of information as its main substance; and all sorts of information weaponry [smart weapons] and systems as its major means.[32]

The adversary would carry out its attacks with non-attributable asymmetric techniques that focus upon information suppression, destruction and alteration, which would be consistent with its doctrine of exploiting the inherent vulnerabilities of information systems. Furthermore, its doctrine would anticipate using highly-trained civilian computer experts as its 'soldiers' in an information war rather than committing large numbers of troops to overrun the ADF. In this, it would seek to deter and even blackmail Australia through the dominance it achieves in possessing information.

As part of the adversary's attempt to corrupt our networks, they would also employ special equipment (both airborne and covertly-emplaced local systems) to intercept the *pro forma* data codes used in our computer-to-computer data exchanges. The *pro forma* include the dial tones of protocols and link-ups that determine the signaling method (such as data transfer multiplexers and private branch exchanges) and the paths and speeds of data transmission.

Buoyed with success against our IT networks, the adversary would also attack our air defences by jamming our radar stations, deceiving the processing, and reconfiguring the displays so that certain azimuths and aircraft types cannot be 'seen'. Adversary aircraft, both manned and unmanned, effectively would have unrestricted access over Australian airspace.

To prevent our air response capability being mobilised, the adversary would inject wireless application protocols (WAPs) and firewall-penetrating software into the avionics of our aircraft as they become airborne, allowing them to be effectively 'hijacked' by that adversary's cyber-specialists with foreknowledge of the details of the hardware and software used in our avionics systems. High-power lasers and radio frequency (RF) weapons would be used to burn out the avionics in any of our aircraft that were not susceptible to 'hijacking'.[33]

Similarly, our maritime response capability would be immobilised by electronic warfare (EW) attacks against the radar systems on our ships and the confounding of ships' communications systems. The radars and other electronic systems aboard every vessel in the fleet would have been carefully studied beforehand—the antenna designs and the signal frequencies, strengths and pulse characteristics—and electronic countermeasures (ECM) equipment and application tactics calibrated to effectively jam or deceive each system. Satellite communications (SATCOM) links with the Australian fleet would be in the adversary's hands, following the 'hijacking' of all of the transponders on our communications satellites.[34]

Critical elements of our command, control and communications system would be destroyed or incapacitated by RF weapons, carried by Unmanned Aerial Vehicles (UAVs) and cruise missiles on one-shot, one-way missions. Ultra-wide band (UWB) weapons, which generate RF radiation over a wide frequency spectrum (nominally from about 100 MHz to more than 1 GHz), but with little directivity, would be used to incapacitate electronic components across broad categories of telecommunications and computerised infrastructure. High-power microwave (HPM) weapons, which generate an RF beam at a very narrow frequency band (in the 100 MHz to 100 GHz range) would be used against our hardened C2 centres, using both 'front door' and 'back door' entry points. In the former case, the RF weapons designers would have used foreknowledge of the antenna systems at the command centres to produce an in-band waveform, with the right frequency and the right modulation to couple with the antennas, allowing the intense energy to burn out components connected to the antennas.[35]

Residual command, control, communications and intelligence (C3I) systems, such as delegated command authorities, fibre-optic cables, rejuvenated sensor systems and makeshift broadcast stations would be dealt with by Special Forces—landed by submarines, and guided by UAVs.[36]

China's cyber-attack capability

As one example of the cyber-attack capability of nation-states, a brief discussion on China is offered here. On 9 December 2007, the *New York Times* reported that in a series of sophisticated attempts against the US nuclear weapons laboratory at Oak Ridge, Tennessee, Chinese hackers had removed data.[37] In March 2007, US Strategic Command Chief General James E. Cartwright told Congress that the United States was under widespread attack in cyber-space.[38] During 2007 and 2008, 12 986 direct assaults on federal agencies and more than 80 000 attempted attacks on Department of Defense computer network systems were reported. Some of these attacks reduced the US military's operational capabilities.[39]

An American cyber-security company that focuses on centralised activity emanating from China reported that attacks from China had almost tripled in the three months before December 2007.[40] In December 2007, Jonathan Evans, the chief of the United Kingdom's domestic counter-intelligence agency MI5, warned a number of accountants, legal firms, and chief executives and security chiefs at banks that they were being spied on electronically by Chinese state organisations. Evans noted that a number of British companies, with Rolls Royce being one example, had discovered that viruses of Chinese Government origin were uploading vast quantities of industrial secrets to Internet servers in China.[41]

Earlier, in October, one of Germany's top internal security officers, Hans Elmar Remberg, told a Berlin conference on industrial espionage that his country was involved in a cyber-war with the Chinese, arguing that he believed Chinese interests were behind the recent digital attacks. One month earlier, in September, the French Secretary General for National Defence, François Delon, argued that he had proof that there was involvement from China in cyber-attacks on France, but could not conclusively say it was the Chinese Government.[42]

The German Government too has been subjected to cyber-attacks, with German Chancellor, Angela Merkel, being informed in August 2007 that three computer networks in her own office had been penetrated by Chinese intelligence services. A few days later, she confronted the visiting Chinese premier directly about the attacks. Premier Wen Jiabao was shocked and promised that his government would get to the bottom of the matter. Interestingly, he then asked for detailed information from Germany's counter-intelligence agencies to help China's security police find the culprit.[43]

Notwithstanding the wide-ranging nature of such attacks, by far the target attacked most intensely by the Chinese is the US military, closely followed by the State Department, the Commerce Department, and the US Department of

Homeland Security. Computer networks in sensitive US sectors relating to commerce, academia, industry, finance, and energy are also being targeted. Indeed, one US cyber-security expert told a group of federal managers that 'the Chinese are in half of your agencies' systems'.[44]

While the United States and other governments appear reticent to reveal the extent of the vulnerabilities of their databases to Chinese penetration, the information available does tend to indicate how widespread Chinese cyber-attacks have become. Cyber-warfare units in the Chinese People's Liberation Army (PLA) have already penetrated the Pentagon's Non-classified Internet Protocol Router Network (NIPRNet) and have designed software to disable it in time of conflict or confrontation. Major General William Lord, Director of Information, Services, and Integration in the US Air Force's Office of Warfighting Integration, admits that China has downloaded data from the NIPRNet already, and that China now poses a nation-state threat.[45]

Richard Lawless, Deputy Undersecretary of Defense for Asia-Pacific Affairs, told a Congressional panel on 13 June 2007 that the Chinese are 'leveraging information technology expertise available in China's booming economy to make significant strides in cyber-warfare'. He noted the Chinese military's determination to dominate, to a certain degree, the capabilities of the Internet, and that this capacity provides them with a growing and very impressive ability to wage cyber-warfare.[46]

Lawless argued that the Chinese have developed a very sophisticated, broadly-based capability to attack and degrade US computer and Internet systems. He noted that the Chinese well-appreciated how to access computers and disrupt networks, particularly by penetrating the networks to glean protected information and by carrying out computer network attacks, which would allow them to shut down critical systems in times of emergency. Lawless argues that the capability is already there; it is being broadened; and it represents a major component of Chinese asymmetric warfare capability.[47] It is also believed that PLA cyber-warfare units have access to source codes for America's ubiquitous office software, giving them a skeleton key to every networked government, military, business, and private computer in the United States.

As China's cyber-attack capability becomes clearer, the US Government will need to acknowledge the vulnerability of America's national security information infrastructure as well as its commercial, financial and energy information networks and make the requisite changes. And, via their computer network operations, China's clandestine intelligence collection would seem to be the foremost intelligence threat to America's science and technology secrets.[48] Clearly, this applies equally to Australia.

What should we do?

As our computer networks proliferate and dependencies on them increase, their protection against cyber-attacks needs far more attention than previously. The knowledge and abilities of hackers have become much more sophisticated and are outstripping methods that detect, identify and alert users to network attacks. These current cyber-defence methods tend to rely on data mining approaches that are useful for simple attacks but not for the more complex or coordinated attacks of recent times. Cyber-space security urgently requires next-generation network management and Intrusion Detection Systems (IDS).

One method of dealing with the emergent challenges is to address network security from a system control and decision perspective that would combine short-term sensor information and long-term knowledge databases to provide improved decision-support systems as part of improved C2. In this respect, information fusion is needed to provide the foundation for a decision and control framework that can detect and predict multi-stage stealthy cyber-attacks.[49]

The problem with our networks

Large networks today contain a multitude of hardware and software packages and are connected in multiple ways. Inevitably, the complexity of these networks introduces security problems that evade detection by network administrators, who tend to focus on isolated and discrete vulnerabilities.

Cyber-attacks traditionally have been one-dimensional—Denial of Service (DS) attacks, insertion of computer viruses or worms, and unauthorised intrusions (referred to as 'hacking'). These attacks were mainly launched against websites, mail servers or client machines. For the future, cyber-threats will be more diversified and take the form of multi-stage and multi-dimensional attacks that utilise and target a variety of attack tools and technologies. For example, the latest generation of worms uses a variety of different exploits, propagation methods, and payloads to inflict damage. Networks and computers that become infected are used to launch attacks against other networks and computers and may access or delete data held within them.

Framework for network defence

In improving cyber-defence, the awareness of potential attacks and an assessment of the impacts of those attacks must be made. This is typical risk assessment/risk management. Recent advances in applying data fusion techniques offer part of the solution.

The fusion of data allows basic awareness and assessments to be refined in order to identify new cyber-attacks. Recognition of the features of such attacks must be both dynamic and adaptive in order to generate initial estimates of the situation and to respond as new or unknown forms of attack emerge.

The following types of *network-based attacks*, focused on web, email or network attack, were identified as the most likely by Shen et al:[50]

- *Buffer overflow* (web attack)—This occurs when a program does not check to ensure the data it is putting into a buffer area will actually fit into that area. A vulnerability currently exists in Microsoft IIS 5.0 running on Windows 2000 that allows a remote intruder to run arbitrary codes on the targeted machine, whereby the intruder is able to gain complete administrative control of the machine. A remote attacker could send a request that might cause the web server to crash with unexpected results.

- *Semantic URL attack* (web attack)—In a semantic Uniform Resource Locator (URL)[51] attack, a client manually adjusts the parameters of its request by maintaining the URL's syntax but altering its semantic meaning. This attack is primarily used against Common Gateway Interface (CGI)[52] driven websites. A similar attack involving web browser cookies[53] is commonly referred to as cookie poisoning.

- *Email bombing* (email attack)—An email bomb involves sending large volumes of email across the Internet to an address, seeking to overflow the mailbox or overwhelm the server. One possible response to this form of attack is to identify the source of the email bomb or spam and configure the router (either internally or through the network service provider) to prevent incoming packets from that address.

- *Email spam* (email attack)—Spam involves the use of electronic messaging systems to send unsolicited, undesired bulk messages, without the permission of the recipients. The addresses of recipients are obtained from network postings, web pages, databases, or through guesswork, using common names and domains.

- *Malware attachment* (email attack)—Malicious software (or malware) is software designed to infiltrate or damage a computer system without the owner's informed consent. Common Malware attacks include worms, viruses, and 'Trojan horses'.

- *Denial of service* (network attack)—A DS attack is an attempt to make a computer resource unavailable to its intended users. Typically, the targets are high-profile web servers, with the aim of the attack being to make the hosted web pages unavailable on the Internet. A DDS attack occurs when multiple compromised systems flood the bandwidth or resources of a targeted system, usually a web server. Systems can be compromised through a variety of methods and attacks may be multi-stage. For example, email spam and Malware may be used first to gain control of several temporal network nodes, which might be poorly-protected servers, followed by a DS attack that is triggered to a specific target.

From a network defence perspective, the following *defensive actions* were considered by Shen et al:[54]

- *Deployment of Intrusion Detection Systems*—An optimal deployment strategy of IDS should be used to maximise the chance of detecting all possible cyber-network attacks and intrusions.
- *Firewall configuration*—A firewall is an IT security device which is configured to permit or deny data connections that may be set and configured by the organisation's security policy. Firewalls can be either hardware or software based or both.
- *Email-filter configuration*—Email filtering involves organising email according to specified criteria. While this usually refers to the automatic processing of incoming messages, it also applies to outgoing emails and can involve human intervention as well as automatic processing. Email filtering will usually do one of three things—pass the message through unchanged for delivery to the specified user's mailbox, redirect the message for delivery elsewhere, or discard the message. Some email filters are also able to edit messages during processing.
- *Shut down or reset servers*—This should eliminate the threat; however, it also denies access to authorised users for the period of the shut-down.

Policy-makers, government administrators, infrastructure owners and operators need to work together to protect Australia against IW, implementing eight key initiatives.[55] First, all external entities who interface with the Australian Government's information infrastructure should be required to be ISO 17799 certified[56] or a new standard yet to be developed.

The ISO standards cover the following 12 domains: risk assessment and treatment, security policy, organisation of information security, asset management, human resources security, physical and environmental security, communications and operations management, access control, information systems acquisition, development and maintenance, information security incident management, and business continuity management and compliance. Security analysts responsible for protecting sensitive information can be relatively confident if an organisation is ISO 17799 compliant. A high degree of information assurance (security) is likely to be a characteristic of the data set being used. The reverse may be true if the organisation is not compliant with ISO 17799.[57]

Second, the Australian Government needs to oversee an information security awareness training program across all sectors, highlighting the threats and vulnerabilities associated with the use of the information infrastructure (especially computer networks and the Internet).

Third, all sensitive national research and development programs should implement an information security plan that includes personnel screening practices, security training, and monitoring practices. The methods and means used by unfriendly competitors or hostile nation-states and the nature of modern-day information processing technology dictate that we must be vigilant in protecting critical information assets and our national research infrastructure.

Fourth, access to the Internet by Australian Government employees should be restricted and isolated. Only a limited number of employees need to have direct access to the Internet. In such cases, the workstations should be completely isolated from the internal network.

Fifth, communications on all Australian Government information systems should be encrypted.

Sixth, any products, materials, integrated circuits, components, programs, processes or other goods that are deemed to be crucial to national security of Australia should be manufactured exclusively in this country or from highly trusted allies.

Seventh, private owners of information networks that interface with any of the nation's critical infrastructure (e.g. defence, law enforcement, finance, and energy) should be required to become ISO 17799 certified. Critical infrastructure is defined by Perry as 'services that are so vital that their incapacity or destruction would have a debilitating impact on the defence or economic security of Australia'.[58]

Eighth, Australia should do all it can to produce more domestic engineers and scientists. We are not growing the intellectual resources that are needed.

Cyber-crime[59]

As just mentioned, the accelerated use of the Internet has led to a dramatic rise in criminal activity that exploits this interconnectivity for illicit financial gain and other malicious purposes, such as Internet fraud, child exploitation, and identity theft. Efforts to address cyber-crime[60] include activities associated with protecting networks and information, detecting criminal activity, investigating crime, and prosecuting criminals.

The annual loss due to computer crime was estimated to be US$67.2 billion for US organisations, according to a 2005 Federal Bureau of Investigation (FBI) survey. The estimated losses associated with particular crimes include $49.3 billion in 2006 for identity theft[61] and US$1 billion annually[62] due to phishing.[63] These projected losses are based on direct and indirect costs that may include actual money stolen, estimated cost of intellectual property stolen, and recovery cost of repairing or replacing damaged networks and equipment.

Cyber-forensic tools and techniques[64] are a key component of cyber-crime investigations as they allow the gathering and examination of electronic evidence that can be useful for prosecution.

Table 2—Techniques used to commit cyber-crimes[65]

Type	Description
Spamming	Sending unsolicited commercial email advertising for products, services, and websites. Spam can also be used as a delivery mechanism for malware and other cyber-threats.
Phishing	A high-tech scam that frequently uses spam or pop-up messages to deceive people into disclosing their credit card numbers, bank account information, Social Security numbers, passwords, or other sensitive information. Internet scammers use email bait to 'phish' for passwords and financial data from the sea of Internet users.
Spoofing	Creating a fraudulent website to mimic an actual, well-known website run by another party. Email spoofing occurs when the sender address and other parts of an email header are altered to appear as though the email originated from a different source. Spoofing hides the origin of an email message.
Pharming	A method used by phishers to deceive users into believing that they are communicating with a legitimate website. Pharming uses a variety of technical methods to redirect a user to a fraudulent or spoofed website when the user types in a legitimate web address. For example, one pharming technique is to redirect users—without their knowledge—to a different website from the one they intended to access. Also, software vulnerabilities may be exploited or malware employed to redirect the user to a fraudulent website when the user types in a legitimate address.
Denial of Service attack	An attack in which one user takes up so much of a shared resource that none of the resource is left for other users. DS attacks compromise the availability of the resource.
Distributed Denial of Service attack	A variant of the DS attack that uses a coordinated attack from a distributed system of computers rather than from a single source. It often makes use of worms to spread to multiple computers that can then attack the target.
Viruses	A program that 'infects' computer files, usually executable programs, by inserting a copy of itself into the file. These copies are usually executed when the infected file is loaded into memory, allowing the virus to infect other files. A virus requires human involvement (usually unwitting) to propagate.
Trojan horse	A computer program that conceals harmful code. It usually masquerades as a useful program that a user would wish to execute.
Worm	An independent computer program that reproduces by copying itself from one system to another across a network. Unlike computer viruses, worms do not require human involvement to propagate.
Malware	Malicious software designed to carry out annoying or harmful actions. Malware often masquerades as useful programs or is embedded into useful programs so that users are induced into activating them. Malware can include viruses, worms, and spyware.
Spyware	Malware installed without the user's knowledge to surreptitiously track and/or transmit data to an unauthorised third party.
Botnet	A network of remotely controlled systems used to coordinate attacks and distribute malware, spam, and phishing scams. Bots (short for 'robots') are programs that are covertly installed on a targeted system allowing an unauthorised user to remotely control the compromised computer for a variety of malicious purposes.

(*Source*: GAO, *CYBERCRIME: Public and Private Entities Face Challenges in Addressing Cyber Threats*, pp. 7–8)

Cyber-crime laws vary across the international community. Australia enacted its *Cybercrime Act* of 2001[66] to address this type of crime in a manner similar to the *U.S. Computer Fraud and Abuse Act* of 1986,[67] which specifies as a crime the knowing unauthorised access to the computers used by a financial institution, by a federal government entity, or for interstate commerce. Such crimes include knowingly accessing a computer without authorisation; damaging a computer by introducing a worm, virus or other attack device; or using unauthorised access to a government, banking, or commerce computer to commit fraud. Violations also include trafficking in passwords for a government computer, a bank computer, or a computer used in interstate or foreign commerce, as well as accessing a computer to commit espionage.

In addition, international agreements are also emerging, such as the Council of Europe's *Convention on Cybercrime* signed by the United States and 29 other countries on 23 November 2001, as a multilateral instrument to address the problems posed by criminal activity on computer networks.

In one example of cyber-crime, a person in the United States was convicted of aggravated identity theft, access device fraud, and conspiracy to commit bank fraud in February 2007. He held over 4300 compromised account numbers and full identity information (i.e. name, address, date of birth, Social Security number, and mother's maiden name) for over 1600 individual victims.[68]

US Department of Defense officials stated that its information network, representing approximately 20 per cent of the entire Internet, receives approximately six million probes/scans a day. Further, representatives from DOD stated that between January 2005 and July 2006, the agency initiated 92 cyber-crime cases, the majority of which involved intrusions or malicious activities directed against its information network.[69]

Indonesian police officials believe the 2002 terrorist bombings in Bali were partially financed through online credit card fraud, according to press reports.[70] As the GAO Report argues:

> As larger amounts of money are transferred through computer systems, as more sensitive economic and commercial information is exchanged electronically, and as the nation's defence and intelligence communities increasingly rely on commercially available IT, the likelihood increases that information attacks will threaten vital national interests.[71]

The effectiveness of the systems put in place to audit and monitor systems, including intrusion detection systems, intrusion protection systems, security event correlation tools, and computer forensics tools,[72] have limitations that impact their ability to detect a crime occurring.[73]

Addressing our critical information infrastructure

When the Estonian authorities began removing a statue of a Soviet soldier (a Second World War memorial) from a park at the end of April 2007, a number of DDS attacks disabled various websites such as the Estonian parliament, banks, ministries, newspapers and broadcasters. The attacks overloaded the bandwidths for the servers running the websites.[74]

These attacks were not initiated by the Russian Government or its security service as was initially suggested. Fake Internet Protocol (IP) addresses were used, which pointed to a Russian Government computer. The attacks were low-tech and probably carried out by large numbers of 'script kiddies'—teenagers with relatively little real computer expertise, who use readily available techniques and programs to search for and exploit weaknesses in computers via the Internet.

This reinforced the concern of many that Critical Infrastructure Protection (CIP) needs greater attention. There are two interlinked and at times reinforcing factors that concern nations: the expansion of the threat spectrum in recent years, especially in terms of malicious actors and their capabilities; and a new kind of vulnerability due to modern society's dependence on inherently insecure information systems.[75]

Both of these factors combine to pose challenges that involve uncertainty over who, how, where, what, why, and when.[76] The notion of an imminent, direct and certain threat, which is how we treat most military threats, does not describe these challenges adequately. They are actually indirect, uncertain and future risks rather than threats.[77] For this reason, the focus needs to be as much on general vulnerabilities of society as it is on actors, capabilities and motivations.

Critical infrastructure is deemed critical because its incapacitation or destruction would have a debilitating impact on the national security and the economic and social welfare of a nation. Examples are telecommunications, power grids, transport and storage of gas and oil, banking and finance, traffic, water supply systems, emergency rescue services, and public administration. Fear of asymmetric measures being perpetrated against such targets has been aggravated by the information revolution.

Today, almost all critical infrastructure relies on various forms of software-based control systems for effective, reliable and continuous operation. Information and communication technologies (ICTs) connect infrastructure systems in such a way as to make them interrelated and interdependent. Thus, CII has joined the lexicon, and includes computers, software, the Internet, satellites and fibre-optics.

CIIs are generally regarded as inherently insecure. Most of the components are developed in the private sector, where the pressure of competition means security does not drive system design. Computer and network vulnerabilities are therefore to be expected, and these lead to information infrastructures with in-built instabilities and critical points of failure.[78]

A relatively small attack on infrastructure can achieve a great impact, thus offering a 'force-multiplier' effect to those carrying out infrastructure attacks.[79] The spread of ICT has facilitated access to the tools for attack, and made the success of an attack more likely. In other words, asymmetric attacks against powerful countries have become that much easier to carry out and more likely to occur.

These risks (or asymmetric threats) are of two types—unstructured and structured. The former is random and relatively limited. It consists of adversaries with restricted funds and organisation and short-term goals, such as individual hackers and crackers as well as small groups of organised criminals. The resources, tools, skills and funding available to the actors are too limited to accomplish a sophisticated attack against critical infrastructure and, more important, the actors lack the motivation to do so. They do it for thrill, prestige or monetary gain.

In contrast, structured threats or risks are considerably more methodical and better supported. Adversaries from this group have extensive funding, organised professional support and access to intelligence products, and long-term strategic goals. Foreign intelligence services, well-organised terrorists, professional hackers involved in IW, larger criminal groups and industrial spies fall into this category.

Unfortunately, there are no clear boundaries between the two categories. Even though an unstructured threat is not usually considered of direct concern to national security, there is a possibility that a structured threat actor could masquerade as an unstructured threat actor, or that structured actors could seek the help of technologically skilled individuals from the other group.

Because of the uncertainty of threat, we need to focus on the risk of an event occurring. In this respect, the important question is not what caused the loss of information integrity, but rather what the possible result and complications may be. A power grid might fail because of a simple operating error without any kind of external influence or sophisticated hacker attack.[80] In all cases, the result is the same: a possible power outage that may set off a cascading effect of successive failures in interlinked systems. Analysing whether a failure was caused by a terrorist, a criminal, simple human error or spontaneous collapse will not help to stop or reduce the effect.

Thus, it would seem to be more appropriate to adopt an 'all hazards' approach, designed for protection efforts irrespective of the nature of the threat, with a focus on the capability to respond to a range of unanticipated events. The key is to create greater resilience, commonly defined as the ability of a system to recover from adversity and either revert to its original state or assume an adjusted state based on new requirements.[81]

Structural approaches, and attempts to prohibit the means of IW altogether or to restrict their availability, are largely not feasible because of the ubiquity and dual-use nature of IT.[82] There are also concerns over military reliance on advanced ICT and the extensive IT infrastructure used to conduct operations, as discussed earlier.

Cyber-crime is considered a menace to the economic prosperity and social stability of all nations that are linked into the GII. All nations therefore have an interest in working together to devise an international regime[83] that will ensure the reliability and survivability of information networks. Again, this is more of a resilience strategy than a threat-focused approach. Multilateral conventions on computer crime, such as the Council of Europe's *Convention on Cybercrime* of 2001,[84] could be expanded and built on. International organisations could help develop and promulgate information security standards and disseminate recommendations and guidelines on best practices. International law enforcement institutions and mechanisms, like Interpol, could be used for information exchange—in order to provide early warning of any attack—and cyber-crime investigations. Enhanced cooperative policing mechanisms could also be created.

Comprehensive protection against the entire range of threats and risks at all times is virtually impossible, not only for technical and practical reasons, but also because of the associated costs. What is possible is to focus protective measures on preventive strategies and on trying to minimise the impact of an attack when it occurs.

A key problem currently is that standard procedures do not exist for assessing the risks to critical infrastructure or for recommending security improvements. Furthermore, a framework for agreeing priorities for security remediation of those critical infrastructures deemed the most vulnerable does not exist.

As Stephen Gale (co-chair of the Foreign Policy Research Institute's Center on Terrorism, Counter-Terrorism, and Homeland Security) argued recently,[85] a key initiative for addressing these shortfalls would be the development and deployment of a Security Impact Statement (SIS), analogous to the Environmental Impact Statement (EIS) that exists to protect the environment. Like the EIS, the SIS would be designed as a means for both identifying vulnerabilities and determining the standards and methods to be used in protection and remediation.

The system would need to be able to estimate both the likelihood of specific events occurring and the impacts of alternative security measures to deal with those events. It would need to drive the prioritisation process of investing in security improvements. Experts would be used in determining the likelihood of specific threat scenarios and the likely outcomes of such threats. The optimal system would be one that could provide quantitative estimates of risks and vulnerabilities; clear indicators of priorities for investments in security; and standards for making specific, effective, and efficient improvements in security.

Under such a system, organisations would be required to undertake due diligence reviews for protecting infrastructure and be offered, say, tax credits as partial cost relief in recognition of their improvements in minimising risks to critical infrastructure.

Cryptography

The entire computing industry needs to work together to improve security, as businesses continue facing cyber-attacks. Experts believe that hardware, software, and networks can be made much safer by creating a multi-layered solution. Bill Gates has suggested replacing password protections, often too easily defeated by phishing and other forms of low-tech hacking, with an InfoCard—a digital identity that can be stored in the microchip of a smart card and used to access password-protected websites.[86]

Of course Microsoft has a keen interest in promoting more secure computing environments, since its operating systems are routinely the target of virus attacks. Gates noted, however, that the shift away from passwords would likely take as long as four years because it requires the collaboration of numerous vendors.

While the InfoCard technology should be useful for personal data security, large institutions, such as banks, are looking at large-scale defences to tackle Internet scams. By using the very networks that hackers exploit, companies can fight fraud and cyber-crime at different nodes, instead of in isolation. For instance, if a cyber-criminal in a third-world country exploits a stolen credit card number and then tries to hide behind a proxy server in New York, that New York IP address could be quickly blacklisted and banks and other organisations immediately notified.

While creating more secure technology requires the coordination of software, networks, and hardware, cryptography is at the heart of it. And while no encryption scheme will be completely foolproof, there is a strong effort underway to address security issues before they become major problems.

Conclusion

This chapter has highlighted the value of information to Australia and the ADF today, and discussed the potential forms of IW that could be used against us. There are certain actions an adversary might take against us and certain things we can do to protect ourselves. And there are cyber-crime activities that need to be addressed, as well as CII aspects. This discussion on cyber-attacks and broad network defence flows neatly into the next two chapters, which examine information infrastructure attack in more detail and how we might best secure the Defence and national information infrastructures in Australia.

ENDNOTES

[1] J.V. McGee, L. Prusak, and P.J. Pyburn, *Managing Information Strategically: Increase your Company's Competitiveness and Efficiency by Using Information as a Strategic Tool*, John Wiley & Sons, New York, 1993, pp. 68–69.

[2] Edward Waltz, *Information Warfare: Principles and Operations*, Artech House Publications, Boston and London, 1998, pp. 54–55.

[3] Alvin and Heidi Toffler, *War and Anti-War: Survival at the Dawn of the 21st Century*, Little Brown, London, 1994, p. 140.

[4] See John Arquilla and David Ronfeldt, *The Advent of Netwar*, RAND Corporation, Santa Monica, CA, 1996; and David F. Ronfeldt and John Arquilla, *Networks and Netwars*, RAND Corporation, Santa Monica, CA, January 2002.

[5] Toffler, *War and Anti-War*, p. 142.

[6] Toffler, *War and Anti-War*, p. 148.

[7] See US Department of Defense, *Joint Doctrine for Information Operations*, Joint Publication 3-13, 9 October 1998, available at <http://www.iwar.org.uk/iwar/resources/us/jp3_13.pdf>, accessed 4 March 2008; and the subsequent *Information Operations*, Joint Publication 3-13, 13 February 2006, available at <http://www.dtic.mil/doctrine/jel/new_pubs/jp3_13.pdf>, accessed 4 March 2008.

[8] Discussed generically in Waltz, *Information Warfare: Principles and Operations*, p. 160.

[9] David J. Lonsdale, *The Nature of War in the Information Age: Clausewitzian Future*, Frank Cass, London and New York, 2004, pp. 91–92.

[10] Amy Sands, 'Integrating Open Sources into Transnational Threat Assessments', in Jennifer E. Sims and Burton Gerber, *Transforming U.S. Intelligence*, Georgetown University Press, Washington, DC, 2005, p. 64.

[11] Richard A. Best, Jr. and Alfred Cumming, *Open Source Intelligence (OSINT): Issues for Congress*, CRS Report for Congress, Congressional Research Service, 5 December 2007, p. CRS-1, available at <http://www.fas.org/sgp/crs/intel/ RL34270.pdf>, accessed 4 March 2008. Richard Best is a specialist in National Defense and Alfred Cumming is a specialist in Intelligence and National Security; both are within the Foreign Affairs, Defense, and Trade Division of CRS.

[12] See Mark M. Lowenthal, *Intelligence, From Secrets to Policy*, Second Edition, CQ Press, Washington, DC, 2003, p. 79.

[13] Sands, 'Integrating Open Sources into Transnational Threat Assessments', pp. 64–65.

[14] Sands, 'Integrating Open Sources into Transnational Threat Assessments', p. 65.

[15] Hamilton Bean, 'The DNI's Open Source Center: An Organizational Communication Perspective', *International Journal of Intelligence and Counterintelligence*, vol. 20, no. 2, Summer 2007, pp. 240–57; and Robert K. Ackerman, 'Intelligence Center Mines Open Sources', *Signal*, March 2006, available at <http://www.afcea.org/signal/articles/templates/SIGNAL_Article_Template.asp?articleid=1102& zoneid=31>, accessed 4 March 2008.

[16] This is discussed further in Lonsdale, *The Nature of War in the Information Age*, p. 73.

[17] Ralph Bennett, *Behind the Battle: Intelligence in the War with Germany 1939-1945*, Pimlico, London, 1999, p. 9.

[18] Ajay Singh, 'Time: The New Dimension in War', *Joint Force Quarterly*, no. 10, Winter 1995–96, pp. 56–61(60), available at <http://www.dtic.mil/doctrine/jel/jfq_pubs/1510.pdf>, accessed 4 March 2008.

[19] Waltz, *Information Warfare: Principles and Operations*, pp. 6–7. Waltz also explains on page 27 how IW applies to the physical, information and cognitive domains viz:

- physical—destruction or theft of computers and destruction of facilities, databases, communications nodes or lines;
- information—electronic attack against information content or processes either in the network or during transmission; and
- cognitive—targeted attacks against the human mind via electronic, printed or oral means.

[20] Samuel B. Griffith, *Sun Tzu: The Art of War*, Oxford University Press, London, Oxford, New York, 1963, p. 77.

[21] Colonel R. Szafranski, US Air Force, 'A Theory of Information Warfare: Preparing for 2020', *Airpower Journal*, vol. 9, no. 1, Spring 1995, available at <http://www.iwar.org.uk/iwar/resources/ airchronicles/szfran.htm>, accessed 4 March 2008.

[22] Waltz, *Information Warfare: Principles and Operations*, p. 10.

[23] Waltz, *Information Warfare: Principles and Operations*, p. 18; and Martin Libicki, *What is Information Warfare?*, Center for Advanced Concepts and Technology, National Defense University, Washington, DC, 1995, p. 7.

[24] Waltz, *Information Warfare: Principles and Operations*, p. 28.

[25] Waltz, *Information Warfare: Principles and Operations*, p. 22.

[26] David S. Alberts, John J. Garstka, Richard E. Hayes, David A. Signori, *Understanding Information Age Warfare*, CCRP Publication Series, Washington, DC, August 2001, p. 4, available at <http://www.dodccrp.org/files/ Alberts_UIAW.pdf>, accessed 4 March 2008.

[27] See William G. Perry, *Information Warfare: An Emerging and Preferred Tool of the People's Republic of China*, Occasional Papers Series, no. 28, The Center for Security Policy, Washington, DC, October 2007, available at <http://www.centerforsecuritypolicy.org/modules/newsmanager/center%20publication%20pdfs/perry%20china%20iw.pdf>, accessed 4 March 2008.

[28] This is the stated intention of Chinese IW. See Toshi Yoshihara, *Chinese Information Warfare: A Phantom Menace or Emerging Threat?*, Strategic Studies Institute, U.S. Army War College, Carlisle, PA, November 2001, available at <http://www.strategicstudiesinstitute.army.mil/pdffiles/PUB62.pdf>, accessed 4 March 2008.

[29] Perry, *Information Warfare: An Emerging and Preferred Tool of the People's Republic of China*.

[30] Adapted from Wendell Minnick, 'Computer Attacks from China leave many questions', *Defense News*, 13 August 2007, available at <http://www.taiwanmilitary.org/phpBB2/viewtopic.php?p=38438&sid=8f527c809bde63b7c174fd9b3fbdb7dd>, accessed 4 March 2008.

[31] In the cyber-attack on Estonia in May 2007, some analysts suggest that the network employed may have enlisted more than one million 'botnets'. See Mark Landler and John Markoff, 'In Estonia, What May Be the First Cyberwar', *International Herald Tribune*, 28 May 2007, available at <http://www.iht.com/bin/print.php?id=5901141>, accessed 4 March 2008.

[32] Toshi Yoshihara, *Chinese Information Warfare: A Phantom Menace or Emerging Threat?*

[33] Derived from Gary Waters and Desmond Ball, *Transforming the Australian Defence Force (ADF) for Information Superiority*, Canberra Papers on Strategy and Defence no. 159, Strategic and Defence Studies Centre, The Australian National University, Canberra, 2005, p. 53.

[34] Waters and Ball, *Transforming the Australian Defence Force (ADF) for Information Superiority*, pp. 53–54.

[35] Ira W. Merritt, 'Proliferation and Significance of Radio Frequency Weapons Technology', Statement before the Joint Economic Committee, US Congress, Washington, DC, 25 February 1998, available at <http://www.house.gov/jec/hearings/radio/merritt.htm>, accessed 4 March 2008; David Schriner, 'The Design and Fabrication of a Damaging RF Weapon by "Back Yard" Methods', Statement before the Joint Economic Committee, US Congress, Washington, DC, 25 February 1998, available at <http://www.house.gov/jec/ hearings/02-25-8h.htm>, accessed 4 March 2008; Michael Knights, 'Options for Electronic Attack in the Iraq Scenario', *Jane's Intelligence Review*, December 2002, pp. 52–53; and 'Use It But Don't Lose It', *Aviation Week & Space Technology*, 9 September 2002, p. 29. See also Waters and Ball, *Transforming the Australian Defence Force (ADF) for Information Superiority*, p. 54.

[36] Waters and Ball, *Transforming the Australian Defence Force (ADF) for Information Superiority*, p. 54.

[37] John Markoff, 'China Link Suspected in Lab Hacking', *New York Times*, 9 December 2007, p. A-03, available at <http://www.nytimes.com/2007/12/09/us/ nationalspecial3/09hack.html>, accessed 4 March 2008.

[38] United States Congressional Committee Testimony, 29 March 2007.

[39] See John J. Tkacik, Jr., *Trojan Dragons: China's International Cyber Warriors*, WebMemo no. 1735, The Heritage Foundation, 12 December 2007, available at <http://www.heritage.org/Research/AsiaandthePacific/upload/ wm_1735.pdf>, accessed 4 March 2008. Tkacik also cites a presentation by Dr Andrew Palowitch entitled, 'Cyber Warfare: Viable Component to the National Cyber Security Initiative?' delivered at Georgetown University, Washington, DC, on 27 November 2007.

[40] Stephen Fidler, 'Steep Rise in Hacking Attacks from China', *Financial Times*, 5 December 2007, available at <http://www.ft.com/cms/s/0/c93e3ba2-a361-11dc-b229-0000779fd2ac.html>, accessed 4 March 2008. The source cites Yuval Ben-Itzhak, chief technology officer for Finjan, a Web security group based in San Jose, California.

[41] Rhys Blakely, Jonathan Richards, James Rossiter, and Richard Beeston, 'MI5 Alert on China's Cyberspace Spy Threat', *TimesOnline*, 1 December 2007, available at <http://business.timesonline.co.uk/tol/business/industry_sectors/ technology/article2980250.ece>, accessed 20 December 2007.

[42] 'Now France Comes Under Attack from PRC Hackers', *Agence France Presse*, 9 September 2007, available at <http://www.taipeitimes.com/News/front/archives/ 2007/09/09/2003377917>, accessed 4 March 2008.

[43] John Blau, 'German Gov't PCs Hacked, China Offers to Investigate: China Offers to Help Track Down the Chinese Hackers Who Broke into German Computers', *PC World*, 27 August 2007, available at <http://www.washingtonpost.com/wp-dyn/content/article/2007/08/27/AR2007082700595.html>, accessed 4 March 2008.

[44] Mark A. Kellner, 'China a "Latent Threat, Potential Enemy": Expert', *DefenseNews Weekly*, 4 December 2006, available at <http://www.defensenews.com/story.php?F=2389588&C=america>, accessed 4 March 2008.

[45] 'Red storm rising: DoD's efforts to stave off nation-state cyberattacks begin with China', *Government Computer News*, 21 August 2006, available at <http://www.gcn.com/print/25_25/41716-1.html>, accessed 4 March 2008.

[46] Hearing of the House Armed Services Committee on 'Recent Security Developments In China'; witnesses: Richard P. Lawless, Deputy Undersecretary of Defense For Asia-Pacific Affairs, and Major General Philip M. Breedlove, Vice Director For Strategic Plans and Policy, Joint Chiefs Of Staff; 13 June 2007. Transcript provided by Federal News Service.

[47] Hearing of the House Armed Services Committee on 'Recent Security Developments In China', 13 June 2007.

[48] Tkacik, *Trojan Dragons: China's International Cyber Warriors*.

[49] See Dan Shen, Genshe Chen, Jose B. Cruz, Jr., Erik Blasch, and Martin Kruger, *Game Theoretic Solutions to Cyber Attack and Network Defense Problems*, paper given to 12th ICCRTS Conference, entitled 'Adapting C2 to the 21st Century', 2007, available at <http://www.dodccrp.org/ events/12th_ICCRTS/CD/html/ papers/062.pdf>, accessed 4 March 2008.

[50] Shen, Chen, Cruz, Blasch, and Kruger, *Game Theoretic Solutions to Cyber Attack and Network Defense Problems*.

[51] The URL identifies a resource and its network 'location' so it can be accessed.

[52] The CGI is a standard protocol for interfacing external application software with an information server, commonly a web server. This allows the server to pass requests from a client web browser to the external application. The web server can then return the output from the application to the web browser.

[53] Cookies are parcels of text sent by a server to a web browser and then sent back unchanged by the browser each time it accesses that server. Cookies are used for authenticating, tracking, and maintaining specific information about users.

[54] Shen, Chen, Cruz, Blasch, and Kruger, *Game Theoretic Solutions to Cyber Attack and Network Defense Problems*.

[55] These are extracted from Perry, *Information Warfare: An Emerging and Preferred Tool of the People's Republic of China*.

[56] ISO/IEC 17799 is a set of international information security practices and standards. They specify accepted security practices related to securing information assets. The ISO/IEC 17799 standards (to become ISO 27000 in the future) seek to serve as 'a starting point for developing organization specific (information security) guidelines'. See International Organization for Standardization, *Information technology—Security techniques—Code of practice for information security management*, ISO/IEC, Second edition, Geneva, Switzerland, 16 June 2005.

[57] William G. Perry, 'Enhanced data mining information assurance by using ISO 17799', *Information: Assurance and Security, Data Mining, Intrusion Detection, Information Assurance and Data Networks Security*, Defense & Security Symposium, The International Society for Optical Engineering, 17 April 2006.

[58] Adapted from William G. Perry, 'The Science of Protecting the Nation's Critical Infrastructure', *Voices of Discovery*, Elon University, NC, 7 March 2007.

[59] See US Government Accountability Office (GAO), *CYBERCRIME: Public and Private Entities Face Challenges in Addressing Cyber Threats*, GAO-07-705, Washington, D.C., June 2007, available at <http://www.gao.gov/new.items/ d07705.pdf>, accessed 4 March 2008.

[60] Cyber-crime, as used in the GAO Report, refers to criminal activities that specifically target a computer or network for damage or infiltration and also refers to the use of computers as tools to conduct criminal activity.

[61] Javelin Strategy & Research, *2007 Identity Fraud Survey Report: Identity Fraud is Dropping, Continued Vigilance Necessary*, Pleasanton, CA, February 2007.

[62] US Department of Homeland Security, Remarks by Assistant Secretary Gregory Garcia at the RSA Conference on IT and Communications Security, San Francisco, CA, 8 February 2007, available at <http://www.dhs.gov/xnews/ speeches/sp_1171386545551.shtm>, accessed 4 March 2008.

[63] Identity theft is the wrongful obtaining and using of another person's identifying information in some way that involves fraud or deception. Phishing is a high-tech scam that frequently uses unsolicited messages to deceive people into disclosing their financial and/or personal identity information.

[64] Cyber-forensics employs electronic tools to extract data from computer media storage without altering the data retrieved. Cyber-forensics techniques may also require the reconstruction of media to retrieve digital evidence after attempts to hide, disguise, or destroy it.

[65] GAO, *CYBERCRIME: Public and Private Entities Face Challenges in Addressing Cyber Threats*, pp. 7–8.

[66] Australian Government, *Cybercrime Act: An Act to amend the law relating to computer offences, and other purposes*, No. 161, Canberra, 2001, available at <http://scaleplus.law.gov.au/ html/comact/11/6458/pdf/161of2001.pdf>, accessed 11 March 2008.

[67] The *Computer Fraud and Abuse Act* (18 USC 1030), passed by the US Congress in 1986, was subsequently amended in 1994 and 1996, and again in 2001 by reference to the *Uniting and Strengthening America by Providing Appropriate Tools Required to Intercept and Obstruct Terrorism (USA PATRIOT) Act*, 2001 (Public Law 107-56). The *Computer Fraud and Abuse Act* was itself an amendment to somewhat limited *Counterfeit Access Device and Computer Fraud and Abuse Act*, 1984 (Public Law 99-474), which was the first comprehensive US legislation to identify and provide for the prosecution of crimes committed through and against computer systems.

[68] Statement of Ronald J. Tenpas, Associate Deputy Attorney General before the Subcommittee on Terrorism, Technology and Homeland Security the Committee on the Judiciary, 21 March 2007, available at <http://judiciary.senate.gov/testimony. cfm?id=2582&wit_id=6194>, accessed 4 March 2008.

[69] GAO, *CYBERCRIME: Public and Private Entities Face Challenges in Addressing Cyber Threats*, p. 20.

[70] Alan Sipress, 'An Indonesian's Prison Memoir Takes Holy War Into Cyberspace', *Washington Post*, 14 December 2004, p. A19, available at <http://www.washingtonpost.com/ wp-dyn/articles/A62095-2004Dec13.html>, accessed 4 March 2008.

[71] GAO, *CYBERCRIME: Public and Private Entities Face Challenges in Addressing Cyber Threats*, p. 22.

[72] Intrusion detection systems detect inappropriate, incorrect, or anomalous activity on a network or computer system. Intrusion prevention systems build on intrusion detection systems to detect attacks on a network and take action to prevent them from being successful. Security event correlation tools monitor and document actions on network devices and analyse the actions to determine if an attack is ongoing or has occurred. Computer forensic tools identify, preserve, extract, and document computer-based evidence.

[73] GAO, *Technology Assessment: Cybersecurity for Critical Infrastructure Protection*, GAO-04-321, Washington, DC, 28 May 2004, available at <http://www.gao.gov/new.items/d04321.pdf>, accessed 4 March 2008.

[74] See Myriam Dunn Cavelty, 'Critical information infrastructure: vulnerabilities, threats and responses', in *Disarmament Forum* (Three), 2007, pp. 15–22. Myriam Dunn Cavelty is head of the New Risks Research Unit at the Center for Security Studies at ETH Zurich, Switzerland and coordinator of the Crisis and Risk Network, available at <http://se2.isn.ch/serviceengine/ FileContent?serviceID=CRN&fileid=20009CBA-C36C-C7AC- D7C0-5E43B2974BC5&lng=en>, accessed 4 March 2008.

[75] Cavelty, 'Critical information infrastructure: vulnerabilities, threats and responses', p. 16.

[76] Emily O. Goldman, 'New Threats, New Identities, and New Ways of War: The Sources of Change in National Security Doctrine', *Journal of Strategic Studies*, vol. 24, no. 2, 2001, pp. 43–76 (45).

[77] J. van Loon, 'Virtual Risks in an Age of Cybernetic Reproduction', in B. Adam, U. Beck and J. van Loon (eds), *The Risk Society and Beyond: Critical Issues for Social Theory*, Sage, London, 2000, pp. 165–82.

[78] Michael Näf, 'Ubiquitous Insecurity? How to "Hack" IT Systems', *Information & Security: An International Journal*, no. 7, 2001, pp. 104–18, available at <http://se1.isn.ch/serviceengine/ FileContent?serviceID=PublishingHouse&fileid= 9F1EA165-76C6-BF34-7522-6D4EA03FB0F5&lng=en>, accessed 4 March 2008.

[79] Government of Canada, Office of Critical Infrastructure Protection and Emergency Preparedness, Threat Analysis No. TA03-001, 12 March 2003, available at <http://www.ocipep-bpiepc.gc.ca/ opsprods/other/TA03-001_e.pdf>, accessed 4 March 2008.

[80] Of greater concern than simply hackers is 'hacktivism'. This is the blend of hacking and activism, and describes operations that use hacking techniques against a target's Internet site with the intent of disrupting normal operations but not causing serious damage. Examples are web 'sit-ins' and virtual blockades, automated email bombs, web hacks, computer break-ins, and computer viruses and worms. See Dorothy E. Denning, 'Activism, Hacktivism, and Cyberterrorism: The Internet as a Tool for Influencing Foreign Policy', in J. Arquilla and D. Ronfeldt (eds), *Networks and Netwars: The Future of Terror, Crime, and Militancy*, RAND Corporation, Santa Monica, CA, 2001, pp. 239–88, available at <http://www.rand.org/pubs/ monograph_reports/MR1382/MR1382.ch8.pdf>, accessed 4 March 2008.

[81] John A. McCarthy, 'Introduction: From Protection to Resilience: Injecting 'Moxie' into the Infrastructure Security Continuum', in *Critical Thinking: Moving from Infrastructure Protection to Infrastructure Resilience*, CIP Program Discussion Paper Series, George Mason University, Washington, DC, 2007, pp. 2–3, available at <cipp.gmu.edu/archive/CIPP_Resilience_Series_Monograph.pdf>, accessed 4 March 2008.

[82] Heinrich Böll Stiftung, *Perspectives for Peace Policy in the Age of Computer Network Attacks*, Conference Proceedings, 2001, available at <http://www.boell.de/downloads/medien/DokuNr20.pdf>, accessed 4 March 2008; and Dorothy E. Denning, *Obstacles and Options for Cyber Arms Controls*, paper presented at Arms Control in Cyberspace Conference, Heinrich Böll Foundation, Berlin, 29–30 June 2001, available at <http://www.cs.georgetown.edu/~denning/ infosec/berlin.doc>, accessed 4 March 2008.

[83] A regime can be defined as 'sets of implicit or explicit principles, norms, rules, and decision-making procedures around which actors' expectations converge in a given area of international relations'. See Stephen D. Krasner (ed.), *International Regimes*, Cornell University Press, Ithaca, New York, 1983, p. 2.

[84] Council of Europe Treaty Office, *Convention on Cybercrime*, CETS No. 185, opened for signature in Budapest, Hungary on 23 November 2001, available at <http://conventions.coe.int/ Treaty/Commun/QueVoulezVous.asp?NT=185&CM=8&DF=16/04/04&CL=ENG>, accessed 11 March 2008.

[85] See Stephen Gale, *Protecting Critical Infrastructure*, Foreign Policy Research Institute, November 2007, available at <http://www.fpri.org/enotes/ 200711.gale.infrastructure.html>, accessed 4 March 2008.

[86] Kate Greene, 'Calling Cryptographers', *MIT Technology Review*, 16 February 2006, available at <http://www.technologyreview.com/Infotech/ 16347/?a=f>, accessed 4 March 2008.

Chapter 4

Targeting Information Infrastructures

Ian Dudgeon

Introduction

The national, defence and global information infrastructures underpin and enable today's information society. They play a critical role in how we and others live, and they shape and influence our decision cycle, i.e. what we see, think, decide and how we act. In defence terms, these infrastructures largely determine the functional efficiency of a country's Command, Control, Communications, Computers, Intelligence, Surveillance, Reconnaissance and Electronic Warfare (C4ISREW) and net-warfare capability. And in both defence and broader national security terms, they provide a pathway to psychological operations. Foreign information infrastructures can be targeted to weaken the military capability and national morale of an adversary, and strengthen those of allies and friends. In certain non-war circumstances, foreign infrastructures may also be targeted to project national power, and shape events to national advantage. The ability to target foreign infrastructures for military advantage in combat operations, and to influence the morale and decision-making of friends and foes alike, is recognised as an essential inclusion in the twenty-first century inventory of national capabilities. Australia must develop this capability to protect and project its national interests.

The information society

The advent of the Information Age, now into its second decade, has brought about very fundamental changes in the way that all societies now function.

The information society, including the various information-based environments within, is now a reality. The exponential growth in Information Technology (IT), the accompanying seamless and virtually instant access to vastly disparate information resources, and the flexibility to positively exploit that technology and information across government, business and society generally, is unprecedented.

In all modern functioning nations, which include Australia and all developed and most developing countries, access to and the benefits derived from the information society are taken for granted.

At the personal or micro level, the expectation and ability to instantly communicate and access comprehensive information resources, including domestic and international media broadcasts, iPod downloads, chat rooms or the like, on a 24/7 basis, by such now-basic means as mobile phones,[1] the Internet, and Personal Digital Assistants (PDAs), is the norm.

At the macro level, the provision of reliable information and information technology-dependent services across government and the private sector, is also assumed. Examples include key industries and groupings which are essential to the efficient functionality of the state, such as communications; banking and finance; transportation and distribution services; energy including electricity, oil and gas; water supply; education; the media and other information services; and other essential government services including defence, and emergency services. It is these industries and groupings that enable such activities as share market trading; domestic and international banking; salary and social security payments; electricity distribution; robotics in manufacturing; integrated logistic systems; suburban and interstate train services; air traffic control; all telephone and Internet services; supermarket checkout transactions; and functional C4ISREW in support of the Defence forces.

In less-developed countries, including, for example, some nations in the Asia-Pacific region, the availability of many of the above services across society generally may be limited, because of less infrastructure-related investment or the skills to maintain a reliable service. However, they do exist and generally work well where the local government and key private sector organisations want them to; for instance, for political, defence and other security reasons, and for commercial profitability.

The benefits enjoyed by the information society identifiably include access to better services, enhanced government and private sector capabilities generally and other lifestyle improvements. But the information now available, and the sophistication of supporting technology, impacts across society both quantitatively and qualitatively, and with an intensity, never before experienced. This impact affects not only how we live, but also shapes and influences our decision cycle (what we see, hear, think, decide, and how we act).[2]

For this combination of reasons, information and its supporting technology would—indeed must—routinely be included on the targeting list of any nation at war, especially any nation that includes Information Operations (IO)[3] as a core capability of its warfighting inventory. Furthermore, they would also be included on the list of potential targets of a nation, non-state organisation or individual who seeks to shape or influence specific physical events, or decision-making, across any part of the public, private or general sectors of the information society, in non-war circumstances.

Information Infrastructures: the NII, GII and DII

What is being targeted, by us or by others? The answer is three mostly interconnected and interdependent information infrastructures. The first is the National Information Infrastructure (NII), this being the key network element within a country that enables its information society to function, and determines the efficiency of its functionality. The second is the Global Information Infrastructure (GII), which provides the international connectivity to the NII. The third is the Defence Information Infrastructure (DII), which, as the name implies, serves a country's Defence organisation, both military and civilian.

Definitions of the above vary between authorities, authors and so on within and between countries, but all boil down to the same essential characteristics. ADF definitions have been used below as the primary definitions for the NII and DII because of the ADF's lead role in Australia for targeting any foreign infrastructures. Another definition of the NII, drawn from Defence sources, has also been cited because of its simplicity. Interestingly, the ADF has no definition of the GII.

The National Information Infrastructure

The NII is defined in Australian Defence Doctrine Publication (ADDP) 3-13—*Information Operations* (2006), as

> compris[ing] the nation wide telecommunications networks, computers, databases and electronic systems; it includes the Internet, the public switched networks, public and private networks, cable and wireless, and satellite telecommunications. The NII includes the information resident in networks and systems, the applications and software that allows users to manipulate, organise and digest the information; the value added services; network standards and protocols; encryption processes; and importantly the people who create information, develop applications and services, conduct facilities, and train others to utilise its potential.[4]

The above is more a statement of the composition of the NII than a typical definition, but is very useful in that regard. A shorter definition initially used by Australia's Defence Signals Directorate (DSD) in 1997, in Australia's first national study of the threats to and vulnerabilities of the NII, reads:

> [The NII] comprises those components that make up the network within and over which information is stored, processed and transported. It includes those people who manage and serve the infrastructure, and the information itself. This information may take the form of electronic voice, facsimile or data.[5]

The Global Information Infrastructure

Although there is presently no ADF definition of the GII, the current US Department of Defense definition is

> the worldwide interconnection of communications networks, computers, databases, and consumer electronics that make vast amounts of information available to users. The global information infrastructure encompasses a wide range of equipment, including cameras, scanners, keyboards, facsimile machines, computers, switches, compact discs, video and audio tape, cable, wire, satellites, fibre optic-optic transmission lines, networks of all types, televisions, monitors, printers, and much more. The friendly and adversary personnel who make decisions and handle the transmitted information constitute a critical component of the global information infrastructure. Also known as the GII.[6]

This definition is essentially the same as the US definition for the NII, the only difference being the substitution of the words 'the worldwide interconnection' with 'the nationwide interconnection' as the lead.[7] As for the ADF definition of the NII, the US NII and GII definitions contain a useful description of the composition of components within these infrastructures.

Alternatively, the GII may be defined simplistically as comprising *a global network of NIIs as well as other dedicated international information networks.*

However, the GII, as defined, is not identical with the Internet. The Internet is the global network of networks; other dedicated networks that are stand-alone and not networked, are not part of the Internet.

The Defence Information Infrastructure

The DII is defined in ADDP 3-13 as

> the shared or interconnected system of telecommunications networks, computers, data bases and electronic systems serving the Defence Department's national and global information needs. It is a subset of and comprises the NII, and includes the people who manage and serve the infrastructure, and the information itself. It includes information infrastructure which is not owned, controlled, managed or administered by Defence.[8]

The issue of ownership, control and management applies to the NII and GII as well as the DII, and is discussed further below.

Information Infrastructures: Some key characteristics

A number of key characteristics of information infrastructures flow from the above definitions that are important to targeting considerations. These include *components, connectivity, bandwidth, functional interdependence,* and *ownership and control.*

Components

The NII, DII and GII comprise five distinct interdependent components. The first four are explicit in the definitions above; the fifth is more implicit:

- the *hardware*, e.g. the computers; sensors; physical transmission components such as cable; radio/wireless; satellites, and transmission towers;
- the *software* applications, e.g. processes; protocols; encryption; and firewalls;
- the *information* itself, e.g. the databases; and information in transmission including voice, facsimile, text messages, imagery, or information in other forms;
- the *people* who operate and maintain the infrastructure; and
- *power supply*, without which hardware and software cannot function and information cannot be transmitted or accessed. While integrated backup power supply (e.g. uninterrupted power systems (UPS)) could be considered a part of the hardware component, mains supply is not. Most UPS have only a limited capability in terms of both duration and capacity, and mains supply remains critical for full and enduring functionality.

Connectivity

The very broad, virtually instantaneous and seamless connectivity and reach across the various domestic and international information domains of the NII/GII/DII networks is a characteristic that also contributes significantly to infrastructure functional efficiency. Users of these infrastructures have adjusted business or other practices accordingly. Real-time communications are now critical in many areas of business and government. The domestic and global marketplace, which includes stock exchange and credit card transactions, are such examples.[9] This real-time dependence also applies to many emergency services and especially to Defence functions across the whole C4ISREW spectrum, including sensor to weapon configurations, during combat operations. Disruption to connectivity, even for relatively brief periods of time, could have a major impact on outcomes.

Bandwidth

Bandwidth across all three infrastructures is constantly increasing, particularly over data networks, in parallel with technology improvements. Client demand has not only kept pace with bandwidth availability, but has outstripped it. Broad bandwidth allows access to vast quantities of information in a very short space of time. In a Defence context in particular, it is an important feature of real-time delivery of surveillance and reconnaissance imagery, and the immediate 'pull-down' accessibility of deployed combat forces to their headquarters' intelligence databases. It is also important in emergency services scenarios, for example in real-time or near real-time monitoring of bushfires or other natural disasters where lives are at risk and the timely delivery of humanitarian aid is critical.

Functional interdependence

Functional interdependence between information and its supporting systems, and between the supporting systems themselves, is a major factor related to the functional efficiency and security of any information infrastructure. And the more complex the system or network, the greater that interdependence. Failure, in whole or by a part of any component of an interdependent system, can impact on the functionality of another part or, potentially, on the whole system. Depending on the type of system affected (for example, its size or complexity) and the scale of the failure, the cascade effect can have significant implications for specific or general services and capabilities, and ultimately affect how people live and behave. In military terms especially, this cascade of 'knock-on' effects fits the classic mould of targeting outcomes in 'effects-based' operations.[10]

The principle of related 'effects-based' considerations also applies to any compromise of the five key criteria of Information Assurance (IA), which is discussed in more detail later in this chapter.

Ownership and control

Ownership of the networks that make up the NII, DII (and thus GII) varies between the government and private sector, depending on the country, and what part of the network within that country, is involved. In most countries today, the major telecommunications service providers are privately owned. And in the world of globalisation, those services may be owned, or part owned by foreign private corporations, the exceptions being where the major telecommunications service providers are state-owned enterprises, like, for example, in North Korea.

In addition, the majority of software systems, especially commercial off-the-shelf (COTS) operating systems, are sourced from foreign corporations, as are many specialist hardware components used within those systems.[11]

Furthermore, the people who develop and maintain and administer particular systems within networks, or the networks themselves, will usually be from the private sector, and indeed may be foreigners.

The up-side of the above is that globalisation or selective global marketing enables access, potentially, to the best hardware, software and people services that are available to deliver and maintain key parts of a country's NII and DII.

The down-side is that a country (i.e. its people and government as both shareholders and stakeholders) may not own or, in reality, fully control or manage these vital national services. Potentially at least, systems and networks could be vulnerable (in the production, operating and administrative phases) to a hostile person (local or foreign), acting on behalf of others or alone, who accesses, manipulates (or establishes the means to access and manipulate) the functionality of the systems and networks at a future date. There are many ways how this might be done, within and without the target country, but could include covertly inserting trap-doors and 'Trojan horses' in its operating software, causing the malfunction of key hardware and software components at a critical time, or enabling other hostiles to access exploitable parts of the system. In normal circumstances this situation might not pose a significant risk, but it could become a national security issue in a time of crisis or war.

The percentage of the DII that is made up of and dependent on the NII, and GII, varies from country to country, but it is generally assessed in most technically advanced countries as about 90 per cent or more. Thus, only 10 per cent or less of the DII in these countries falls into the category of being owned, controlled, managed or administered by their Defence organisation. And, generally, the infrastructure that they do own, control and manage exists primarily at the tactical level only. Few countries can afford to have their own fully independent strategic and/or operational broadband communications systems. The United States is one such country, but that resource is nevertheless limited relative to the total size and operational commitments of the US armed forces. This limitation necessitates the majority of communications being transmitted over leased or other non-Defence owned networks.

One important conclusion is, therefore, that a significant proportion of any Defence organisation's C4ISREW capability, its Network Centric Warfare (NCW) 'information superiority' capability, and any potential information-based Asymmetric Warfare capability, is outside its total control, and may well be foreign owned and under actual or de facto foreign control.

It also means that maintaining and protecting a functional and efficient NII (inclusive of DII components) and its GII connectivity is a critical *Defence-specific security requirement*, as much as a broader *national security requirement*.

The Importance of Information Assurance

IA is another key consideration of NII/DII and GII targeting. Effective IA is a critical element in the information society, and underpins both the functionality and efficiency of all information infrastructures.

IA comprises five essential criteria for the protection of information and friendly systems against unauthorised access: *availability*, *integrity*, *confidentiality*, *authentication* and *non-repudiation*:

- *availability* applies to the information itself, its supporting technology and the people who operate and serve the infrastructure;
- *integrity* refers to the trustworthiness of information and system/process reliability;
- *confidentiality* is about denying access to the information and sensitive aspects of supporting technology, to those persons without authorisation;
- *authentication* refers to assuring that those who do access the information or supporting systems have the requisite authorisation; and
- *non-repudiation* is linked to authentication and, effectively, is the digital signature.

The principle that applies to functionally-interdependent systems, whereby the failure of one component can impact on the functionality of one or more other components, also applies to IA. Thus, if any of the above IA criteria are compromised for any reason, at least some element of information and/or functionality and efficiency of related information infrastructures is also likely to be compromised. The more significant the compromise, particularly in key areas or system choke-points or nodes, the more significant the impact will be on functionally and efficiency. Identifying existing vulnerabilities, or creating vulnerabilities that will enable IA to be compromised, is an important part of the targeting process.

The effective implementation of IA involves a wide range of security processes and procedures, as well as physical measures. One important measure is redundancy and diversity, which is intended to counteract the effects of any failure within, or compromise of, a system, or at least to minimise those effects. However, the high-end functionality and efficiency of many of the processes, systems, services and capabilities we rely on and take for granted is dependent, or largely dependent, on current-generation hardware and software. For high-tech systems in particular, the rapid changes in technology resulting in increasingly more powerful hardware and software, means that planned redundancy and diversity to provide effective backup and continuity, must also largely keep pace technologically with primary-use hardware and software.

In Australia and other developed and many developing countries, redundancy and diversity across critical infrastructure has been significantly hardened since the 11 September 2001 terrorist attacks on the United States, but at a cost. However, even high-quality redundancy and diversity might struggle to provide a full service if challenged to do so. But where that quality of investment is not made, or made in depth, technologically-dated backup systems may simply be incapable of maintaining even a basic level of services over a short period to meet national or sector needs, if put to the test.

Redundancy and diversity, however robust, must be recognised as part of the IA equation. They must therefore be factored into targeting considerations.

Targeting Information Infrastructures: who and why?

As indicated earlier above, there are three broad groups of people who, potentially, might target information infrastructures: a nation-state or country for national security reasons; a non-state organisation such as a terrorist, criminal, or political or other Issue Motivated Group (IMG) in pursuit of group objectives; or an individual in pursuit of personal objectives.

Motivation, be it for the short, medium or long term and whether for tactical, operational, strategic or other reasons, is important as it will indicate potential targets, irrespective of whether that targeting process and plan is highly sophisticated and focused or, apparently, more simple, random and opportunistic.[12]

Nation-state targeting

Nation-state targeting may occur during war or non-war situations.

War: targeting the adversary

At the national level, the most obvious scenario for one or more countries to target the NII, DII and GII connectivity of another country or countries is when those countries are at war. The overriding political objective in this scenario would be to win the war by destroying the military warfighting capability and the military and civilian logistic and other support capacity of the adversary or adversaries to wage war; and by changing the will of their military, politicians and civilian population from pursuing the war to seeking peace.

Post-war objectives, ideally shaped during the conflict itself, concerning the future profile of the enemy nation or nations, would include the restoration of a politically viable and economically functional state, sympathetic (in the case of Australia) to Australia's national interests.

Operations and activities undertaken during war against an adversary to achieve the above objectives would, of necessity, be directed at both military and civilian targets, and be both overt and covert.[13] In addition, activities across the adversary's NII, DII and relevant GII linkages, that are undertaken to shape and influence outcomes, would employ destruction and degradation of those infrastructures. Deception and psychological measures would also exploit the adversary's information infrastructures, and those in other countries where these also could contribute to denigrating the adversary's capability and will.

In sum, the measures undertaken would encompass the full spectrum of IO including computer network attack, deception and psychological operations.

War: targeting allies, friends and neutrals

Other activities undertaken during war would target allies and friendly countries to boost their morale and commitment to the war effort and post-war objectives. Such activities would be primarily psychological in nature.

Neutrals would also be targeted, in an effort to persuade them to commit to the allied cause, or at least counter any enemy propaganda aimed at persuading them to commit to the axis cause. Such activities would focus primarily on psychological measures. In certain circumstances, however, where a neutral country or business enterprise within is supplying important military or military-support materiél to the adversary, the information infrastructure within the neutral country that is linked to the production or delivery of that materiél, could be targeted to cease or disrupt supply.

Non-war situations or circumstances short of war

In these situations and circumstances a range of scenarios exist that may politically justify targeting another country's information infrastructures. Scenarios include, but are not limited to:

- Intelligence gathering directed at obtaining information from the NII, DII or GII linkages of the targeted country or countries, through activities as communications intercept, computer network exploitation (CNE)[14] and/or human intelligence (HUMINT) operations. Depending on the country, these operations could be enduring and undertaken to meet specific national intelligence requirements, whether political, strategic, military, economic, societal or any combination of these. The intelligence product would feed the national assessment and policy process of the collector, enabling that collector to plan and undertake action to maximise that country's advantage, or minimise any disadvantage, relevant to the target.

- Assisting friendly opposition elements in a target country that is hostile or opposed to our key national security interests, but does not pose the threat of war, to assist the opposition to bring about a change to that government's policy, or indeed change the government itself. Assistance in this scenario would be mostly covert and generally limited to psychological measures and possibly deception, but would not include destruction or degradation of the NII/DII.

- Pre-emptive action against a country with which we are not at war, but where that country poses an imminent threat to us of war or aggression. As in the previous scenario, activities could be directed at assisting opposition elements in the target country to bring about change to that government's policy or of the government itself. Opposition elements in this scenario could include any armed resistance groups, and assistance could incorporate psychological measures and covert attacks directed at destroying or degrading the NII/DII and GII linkages, and deception.

- Pre-emptive action as above, but where the threat of war or aggression is to allies or friendly governments.
- Activities against hostile occupation forces in an allied or friendly country, where those invasion and occupation forces involved a country with whom we are not at war. Activities would be directed at undermining the capability and morale of occupation forces, while boosting the capability and morale of resistance forces, in order to force the ultimate withdrawal or capitulation of the aggressor. Activities would include psychological measures, deception and the destruction and degradation of the aggressor's DII and relevant parts of the occupied country's NII and GII linkages. Activities generally would be covert, but the veneer of deniability may be thin. In these circumstances, psychological and other appropriate activities would also be undertaken against the targeted aggressor within their country and appropriate third countries in support of these objectives.
- Disruption activities against an aggressor or hostile non-government organisation or individual in a third country or countries (e.g. targeting a terrorist organisation, IMG, criminal group or particular individual who might pose a specific but serious security risk), in order to destroy, disrupt or neutralise that group or individual's actions. Depending on the circumstances, activities could include the destruction or degradation of NII and GII linkages, deception, and psychological measures. Activities could be overt and covert, or a combination of both.

In the case of a terrorist organisation, the breadth and intensity of activities could match aspects of those undertaken in wartime.

In all circumstances in an Australian context, the above activities would have the endorsement of government at the highest level, and be coordinated with other activities that would or may be undertaken by government via other agencies of government, including its Defence forces, on a unilateral, bilateral or multilateral basis. Where appropriate, they would also be coordinated with similar concurrent activities undertaken by allies or friends.

Targeting by non-state organisations

Some targeting considerations applicable to three non-state organisations are reviewed briefly below. They are *terrorist organisations*, *criminals*, and *IMGs*.

Terrorist organisations

The targeting of information infrastructures by terrorists is normally aimed at the dissemination of propaganda that promotes the terrorist's political objectives. Objectives include justifying and proselytising their own cause and actions, disseminating propaganda that emphasises the evils, corruption, injustices or misguided thinking of their opponents or enemies (usually, but not exclusively, ideologies, governments, and individuals within government), winning over converts to their cause, and recruiting new activists. Most terrorist organisations are very adept in the use of psychological measures across media outlets or the Internet, and know how to focus their messages to play on the emotions of their target group. And unlike the Western media or Internet sources which will often censor graphic descriptions or images of tragic scenes, including atrocities committed by terrorists, the terrorists themselves will play up graphically any tragic situation caused by, or which they attribute to, the 'enemy' in order to instil high levels of fear and/or hatred of that 'enemy'. Typical examples are images of 'collateral damage' by their 'enemy', involving graphic pictures of bodies of innocent women and children. Other examples of graphic releases have been 'justifiable acts of revenge' or 'justice' in the form of a video of the execution of an 'enemy' agent.[15]

Other forms of the use of information infrastructures by terrorists are illegal activities to finance their operations, counterfeiting, and money laundering. Examples of the former include drug trafficking and credit card fraud. Tamara Makarenko, an international specialist on criminal affairs, claims that these activities are widespread amongst some terrorist groups.[16] Interestingly, an autobiography published in 2004 and written by Imam Samudra, one of the Indonesians involved in the Bali bombings in 2002, includes a chapter entitled 'Hacking. Why Not?' In this chapter Samudra urges fellow Muslim radicals to take the holy war into cyber-space by attacking US computers specifically for the purpose of credit card fraud. Samudra apparently tried, unsuccessfully, to finance the Bali bombings by attempting such fraud himself.[17] Counterfeiting has been used for the production of false identity documents for the use of terrorist members, while money laundering has been used to hide the sources and quantities of any funds.

Most terrorist organisations are well aware of the vulnerability of their members' communications being intercepted when using telephones (mobile or land line) or the Internet, and have adopted procedures to minimise the risk. These include maximising the use of 'clean' phones by the frequent replacement of the phone and Subscriber Identity Module (SIM) card. One technique for clandestine communication over the Internet is the use of 'steganography', whereby secret messages are concealed beneath overt messages—a sort of electronic microdot. Encryption is also used.

The modus operandi of terrorist organisations can vary significantly depending on their objectives and the sophistication of their training. For example, in the Philippines many terrorist activities have involved the destruction of physical infrastructure by explosives or arson, particularly that owned by companies who refuse to bow to extortion and meet terrorist demands for payment of 'revolutionary taxes'. Targets have included information infrastructure hardware, and/or supporting infrastructure, such as communications towers, electricity transmission towers, and electricity sub-stations. They have been selected not because of any focus on information infrastructure as such, but simply because such targets are accessible and incur less risk to the attackers than optional targets. In this sense, they can be described as opportunistic. But information infrastructure can also be deliberately targeted where this fits with terrorist objectives.

Publicly available information indicates that while there is evidence that a number of terrorists use sophisticated computer skills to conduct activities such as identity theft and credit card fraud, there is no evidence of them having, or at least using, hacking or other skills to conduct large-scale theft or extortion. There is, however, no reason to assume terrorists will not learn these skills and target information infrastructures accordingly, when or if it serves their purpose to do so.

Criminals

Criminals target information and supporting infrastructure mostly for reasons of extortion, theft and fraud. Targeting takes two general forms: targeting the computers themselves to illegally obtain the information within; and using computers as the implement of, or to facilitate, the crime.[18]

Targeting in order to access the computers themselves includes such techniques as hacking, the use of malware to obtain passwords by reading keystrokes, or insider help to obtain passwords. It also includes theft of the information within the computer system itself, such as identity information (e.g. credit card or bank account details of individuals, corporations and government).

The use of computers for criminal purposes has grown in parallel with the growth of computer availability and the information society generally. Related criminal activity includes the theft of intellectual property or other forms of commercial or industrial espionage, the misappropriation of money, money laundering, scams to solicit money using fraudulent investment schemes, and embezzlement. Activity also includes the distribution of illegal material such as child pornography, forgery including breaches of copyright (e.g. production of pirated compact discs and software), and extortion in its many forms. Types of extortion include the threat of destroying a corporation's key databases or disclosing confidential information within those databases. Implicit in the blackmail is the willingness to carry out the threat if the corporation fails to meets the extortionist's monetary demands.

Part of the criminal modus operandi has been the use of various skills or methodologies to obtain the necessary personal or other data to access targeted sites. For example they might use hacking, spyware or phishing to obtain passwords or personal information, including bank account or credit card details, for later exploitation. Or they might use a combination of cyber and non-cyber methods to attack a target, for instance using an 'insider' to provide passwords or other operational data necessary to then plan and mount a cyber-attack. The skills and methodologies used, at least by some criminals, are very sophisticated. The use of state-of-the-art software, in operations involving theft and fraud, may take some time to detect and counter. While the above comprises a diverse range of criminal activities using cyber-crime, the general categories of crime are not new: it is, instead, the technology and circumstances involved that are new.

Like terrorists, criminals also communicate across information infrastructures using techniques to avoid detection and interception. Such methods also include steganography, encryption and cut-outs. Indeed, because of the overlap of systems, it is probable that many of the techniques used by terrorists were initially sourced from criminals.

Who the criminals are, and the level of their sophistication, will determine their target selection. International computer crime statistics have shown a steady annual increase in the frequency of cyber-crime, and reports have surfaced occasionally about some allegedly highly sophisticated acts of extortion, theft and fraud involving very significant sums of money.[19] The incidence of computer crime (or cyber-crime) can be expected to increase in proportion to the growing awareness of and exposure to potential opportunities, assisted by increasingly higher levels of computer literacy generally, and accessibility to sophisticated hacking techniques.[20]

The simple conclusion that can be drawn from this situation is that where money (or other gain) is involved, the threat of criminal targeting is inevitable. And the larger the amount of money involved, the more probable the threat.

Issue Motivated Groups

IMGs, including politically motivated groups, most often target information infrastructures to exploit their reach so as to promote issues to their advantage. Such promotion may be information that explains or supports their position, or denigrates that of their opponents. The Internet is frequently used for this purpose and includes the hosting of websites, and the use of other means such as YouTube and chat rooms. Issues can be enduring, such as those supporting a Palestinian homeland or opposing the war in Iraq; more recent issues include promoting protection of the global environment (which is expected to become an enduring issue), and current issues include opposition to Japanese whaling.

However, some IMGs actively seek to hack into websites and access information which they are not authorised to receive, and particularly information which, if leaked, would damage their opponent's position and thus support their cause. Forged documents or other fabricated information that advantages an IMG's position is not that uncommon, and is indeed often used to keep alive or inflame an issue.

Other forms of action by IMGs can include 'electronic vandalism' as a protest against a specific organisation, by defacing a web page or closing down a website by the use of Denial of Service (DS) attacks. However, while some IMGs are a threat to the confidentiality of information that can be exploited to advance their cause, and do target the functionality of specific infrastructure targets by electronic vandalism, in general they are not a threat to the broader functionality of the infrastructure itself.

Individuals

Individuals are a 'wild card'. They vary from the benign to the highly dangerous. The former include gifted hackers, some of whom break into restricted computer systems simply to prove to themselves or their colleagues that they can do it, but take no further action. They are sometimes referred to as 'joyriders'. However, as indicated above, other hackers are highly destructive. They include writers of malicious software programs, or malware, such as viruses and worms that can rapidly replicate themselves across computers and networks and cause significant damage through the destruction or corruption of operating systems and databases. This group of hackers also includes individuals who deliberately destroy or degrade hardware and software, or steal, disclose or enable the disclosure of confidential information from within. They include revenge seekers such as disgruntled employees, clients or customers.

Threats from individuals are a reality, but the specifics of who, what and when remain largely unpredictable.[21]

Targeting: objectives

Having painted the broader background canvas of targeting information infrastructures, this chapter now reviews the issues of *objectives*, *targets*, *capabilities required*, *vulnerability*, *accessibility* and *intelligence*. The context is Defence related, specifically a country at war, but considerations may be scaled down for application to limited war, crisis or other non-war situations.

The starting point for any plan of action is the end-state or objectives to be achieved. As earlier stated, the overriding objective of a country at war would be victory by destroying the military capability and military and civilian support capacity of the adversary to wage war, and to change the will of their military, politicians and civilian population from pursuing the war to seeking peace. Post-war objectives concerning the future profile of the enemy, shaped during the conflict itself, would include a politically and economically viable and functional society sympathetic (in the case of Australia) to Australia's interests. Other objectives included boosting the morale and commitment of allies and friends, and countering efforts by the enemy to win over the sympathy or support of neutrals.

Targets were seen as psychological or cognitive, and 'physical' in terms of destroying, degrading or manipulating hardware, software, information, and power supply. People who operated or maintained the infrastructure could be targeted both psychologically and physically. Activities undertaken also could be overt and covert.

Both psychological and physical targets impact on the decision cycle. While perhaps self evident in the case of psychological operations, anything that can impact on an enemy's collection or 'observe' capability[22] (e.g. reconnaissance and surveillance) or their analysis or 'orient' capability (e.g. access to intelligence databases) will affect decisions and actions. And, often, incorrect or impaired decisions and actions can generate their own psychological effects which may adversely impact on morale and commitment, and on subsequent decision-making.

The above overarching objectives are strategic; objectives would also be set at the operational and tactical level, in all cases in support of and coordinated with achieving the higher aim.

Some examples follow. A basic military capability objective at the strategic/operational/tactical level would be to hide or disguise friendly-force manoeuvre from enemy observation, to achieve surprise. Targeting in this scenario would include the destruction or disruption of critical enemy satellite ground stations in order to neutralise headquarters-controlled imagery, reconnaissance and surveillance resources at that time. This action would also be supported by electronic deception activities.

Another basic objective would be the disruption to an enemy's logistic supply capability that provided critical support to the enemy's combat capability needs. Information systems on which such logistic supply depends could be targeted at the strategic, operational and tactical level. Besides the direct impact of logistic disruption to the enemy's warfighting capability, this impact would also affect enemy morale. Negative morale could be further exploited by appropriately tailored psychological operations (psyops).

A further objective using psyops to reach a broad multinational audience would be targeting the morale of the enemy, the morale and commitment of their supporters, the commitment of our allies and friends, and attitudes of neutrals, by the dissemination of adverse publicity, or propaganda, about any enemy illegal, unethical, immoral, or otherwise culturally insensitive activities. Issues exploited could include those affecting the treatment or welfare of non-combatants in occupied areas, particular religious or ethnic groups, or political or military prisoners.[23]

Methods of delivering intended messages or publicity across information infrastructures could include all forms of domestic and international media (radio, television and newspapers), the Internet, and the use of phones, especially mobile phones (via oral or Short Message Service (SMS)), in enemy country or countries, and appropriate third countries. Methods chosen may need to circumvent severe censorship in enemy and enemy-occupied countries. Scope also exists for the focused use of disinformation. However, while disinformation can be an effective tool, especially for short-term gain, it is a double-edged sword. User credibility can be compromised if it is blatantly exposed or excessively used by the same source.

A final example relates to coordinated cyber-attack and psychological operations at the tactical level against enemy occupation forces in a third country. This scenario assumes our forces are deployed against the enemy in the combat zone, the existence of an armed and capable resistance behind enemy lines with whom we are in contact and can coordinate operations, and whose post-war political ambitions, together with those of the local population generally in enemy-occupied areas, are compatible with ours. The enemy's C4ISREW capability would be fully targeted, using all resources available to us, including the local resistance against related targets as directed by us. The result, especially at a time of proposed major combat operations aimed at inflicting serious losses on and, potentially, the withdrawal or capitulation of the local enemy forces, would be to severely disrupt the enemy's surveillance capability and thus knowledge of the battlefield, his decision-making and C2 capability generally, thereby giving us the significant combat advantage we need to win.

Concurrent psyops, and those mounted after the battle, would be directed at boosting the morale of the resistance and local population generally, and demoralising further the enemy. Psyops would not only target friend and foe in the immediate area of combat operations, but those in adjacent areas that will be the scene of future combat operations.

The above are indicative examples only, and specific objectives may involve highly complex plans necessitating the coordination of highly varied resources across many different countries. But the starting point is clearly identifying the objectives (primary and other), both physical and psychological, and then applying the requirements, assessment and planning methodology.

Targeting: capabilities required

Australia requires a comprehensive physical and psychological operations capability to effectively target, attack or exploit information infrastructures in support of its national and Defence interests. That capability is essential, not just as a part of a twenty-first century warfighting capability, but also as an option for power projection across the whole spectrum of national interests in both war and non-war situations. Developing and maintaining such a capability would not be without challenge.

All significant national security and Defence capabilities cost money and resources, and this chapter does not attempt to cost all resource requirements. Four considerations do, however, apply:

- First, Australia has long recognised the need to develop and maintain a technological edge in its regional warfighting capability. That edge has been progressively eroded as neighbours develop and acquire new technology weapons and thus capabilities. IO is not only a new technology capability, but offers to many countries a significant asymmetrical warfare capability if properly used. All regional countries in East Asia are aware of this. In the case of technologically advanced countries such as China, Japan, South Korea, Taiwan and Singapore, the concept of IO has been accepted, and the capability is reportedly well developed. If Australia is to remain in the advanced-technology sphere, IO is a capability that must be acquired.
- Second, used in the right circumstances and the right way, the application of this holistic capability in some circumstances may well avoid the need to engage in combat operations. Or it might otherwise significantly reduce the scope of combat operations. Not only might this save a significant number of lives of Australian Service personnel, but it would be a significantly cheaper option than engaging in major combat operations. The cost of a comprehensive capability needs to be assessed against the totality of the national security budget, the cost of other warfighting capabilities, their effectiveness at the strategic, operational and tactical levels, and particularly the cost of major combat inventory items such as aircraft and ships that could be lost in combat operations.
- Third, acquiring the capability would not be a substitute for other existing or planned capabilities, but would supplement those capabilities. This new capability offers not only support to existing capabilities, but a greater reach of military and general national security options, and a greater flexibility of options.
- Fourth, the blocks on which to build a comprehensive capability are already in place. What is required is the identification and acquisition of the necessary additional equipment and skill sets, in a similar vein to any other priority national security requirement.

A comprehensive capability to target information infrastructures includes, but is not limited to the following:

Psychological operations

A basic psyops capability exists within Defence. The Department of Foreign Affairs and Trade (DFAT) does not have this specific capability; it has expertise in advocacy. Understanding and developing a capability that properly meets national IO requirements needs to be developed. Part of such a development will necessitate comprehensive databases within the Defence Intelligence Organisation (DIO) on the region's inhabitants (demographics, characteristics of race, culture, religion, relationships within and between ethnic and religious groups, national aspirations, and attitudes to Australia). Databases must also include comprehensive details on how these individuals communicate domestically and internationally (e.g. Internet, media, landline and mobile telephone, personal meetings) and the relative influence of the different methods of communications.

Much of this knowledge already exists within Australia, within various universities, Non-governmental Organisations (NGOs), and our immigrant population. But accurately interpreting this information, and knowing how to properly and effectively use it to meet specific objectives, requires a great deal of skill. It is essential that the people be profiled to accurately reflect who they really are and what attitudes and thinking they hold, rather than who we would like them to be and what we would like them to think.

'Wargaming' in psyops involving, as a minimum, Defence, DFAT, the Department of Prime Minister and Cabinet (DPM&C) and relevant Ministers is also essential. Key decision-makers and those responsible for policy implementation and practice must develop an informed understanding of the most appropriate use of psyops in different circumstances, including nation building and national cohesion.

Database management

A detailed knowledge and maintenance of current databases is required of the NII, DII and GII linkages in all relevant countries within our regional area of interest. Databases should include specific knowledge about operational aspects, vulnerabilities and accessibility relevant to the systems and networks that make up these structures. The DSD will have a lead role in assembling this knowledge.

Computer Network Operations (CNO)

Knowledge and experience is required in the full range of CNO skill sets, including destruction, degradation, manipulation, and intelligence extraction.[24] This must include the practical know-how of hacking into all relevant commercially available and other IT operating systems, and the development and placement of trap-doors, 'Trojan horses', viruses, worms and the like. Knowledge and expertise must also include cascade effects, and how to monitor the effects of CNO. DSD and DIO will have the lead roles in developing this capability.

The potential application of CNO skill sets raises a number legally (and politically) sensitive issues such as hacking through the information infrastructures of neutral countries, and the ability to destroy or disrupt major commercial IT systems or equipment, including satellites, by different forms of electronic attack. These legal issues need to be addressed, and also considered in the Rules of Engagement (ROE) in any operational application.

Other weapons and methodologies

Other weapons and methodologies for attacking and destroying or degrading information systems must be developed. Weapons will include those that employ high-powered microwave, and directed energy, including laser beams. The Defence Science and Technology Organisation (DSTO) will have a lead role in the development of this capability.

Media

It will be necessary to acquire a radio and television capability that can beam broadcasts into target countries. In the event that these broadcasts are jammed by the target country, a capability to break into and override existing radio and television broadcasts within the target country is also required. Psyops specialists, in cooperation with other key contributors, will have a lead role in selecting which individuals should do the broadcasting, and the appropriate content of broadcasts.

HUMINT assets

HUMINT assets must be acquired and other specialist HUMINT assistance provided, in target and third countries, in support of all aspects of physical, psychological and deception operations. The Australian Secret Intelligence Service (ASIS) would have the lead role in providing these assets and assistance.

Additional capabilities

Any other capabilities which are required would build on those already in existence. These include conventional and other specialist intercept techniques, decryption and EW.

Targeting: vulnerability and accessibility

Vulnerability and accessibility are also critical elements to the targeting process. There is an interdependence between both these elements and capability. Any potential vulnerability that is also accessible cannot be exploited unless the attacker has the requisite capability.

There is also an interdependence between vulnerability and accessibility. As for capability, any known or potential vulnerability cannot be exploited unless some relevant component is accessible.

Vulnerabilities

Vulnerabilities include:

- Identified weaknesses in a system due to inadequate security procedures or processes designed to prevent unauthorised access (e.g. passwords, level of encryption).
- Weaknesses due to the failure of a person or persons to follow proper security procedures to prevent unauthorised access (e.g. improper disclosure of passwords or security procedures, disclosure of classified information over open line telephones or the Internet, failure to secure buildings or security containers housing critical hardware or software).
- Physical access to parts of the infrastructure that are not protected by physical or electronic barriers of some kind (e.g. fibre-optic cable runs or radio/microwave transmission towers outside protected establishments).
- Nodes or physical choke points where different parts of an infrastructure are concentrated and which, therefore, offer a rich assembly of targets. Nodes can offer the benefits of economy and concentration of force, and the outcome, if attacked, of more significant damage and delays in restoring functionality than if an individual component only was attacked.
- Vulnerability may be a product of interdependence and complexity, i.e. the more interdependent and complex the infrastructure, the more vulnerable the information or systems if almost any part of the infrastructure is destroyed, disrupted or manipulated.
- Vulnerability may also be a product of the time required to repair the infrastructure or reinstate business continuity, e.g. the longer the time it takes to repair or replace hardware, software or human components to restore functionality, the more vulnerable the target. Critical components that significantly affect functionality and require extended time to repair or replace are the preferred targets.

Accessibility

Accessibility is multifaceted. It may seek to target one or more of the key criteria of IA, e.g. availability, integrity or authentication. It could also target any one component of hardware, software, information, the people who operate and maintain, or power supply, or it may be a combination of these. Targeting might be by direct access to the infrastructure, or indirect access in or from third countries.

Examples of direct access in order to destroy or disrupt key hardware could range from a missile strike, to sabotage by resistance forces or Special Forces. Direct access in order to intercept the enemy's communications may require 'tapping' into accessible fibre-optic cables. It might also include destroying an enemy's primary communications route that is inaccessible to tapping or other forms of intercept, in order to force the target to use an alternative communications route that is accessible. HUMINT assets, potentially, could assist directly in all the above situations.

Indirect access to degrade, corrupt or manipulate data within a critical enemy intelligence or logistic database could be achieved by hacking into that database through third countries. Important information about an enemy's intentions might also reside in, for example, their embassy or an axis partner's embassy in a third country. That information might be accessible in that country, but not elsewhere, through HUMINT, signals intelligence (SIGINT) or CNO operations mounted there.

A final example could be 'disruption' at a critical time to enemy communications across a foreign-owned satellite, through cooperation by the foreign owners/operators of the satellite service, or a HUMINT asset among those who operate or maintain that facility.

All information infrastructures are potentially vulnerable in some way. The issue is where and how. While an initial assessment might suggest a particular objective is impossible, it might actually prove to be possible with the application of lateral thinking. The issue then is whether the risk and resources are worth it.

Intelligence

The quality of intelligence input into the targeting of information infrastructures will largely determine the effectiveness of the targeting outcomes.

As previously mentioned, the intelligence applies to psychological requirements as much as to requirements related to all components of the infrastructures themselves. These requirements are comprehensive and need to be identified and fulfilled in advance of any crisis or conflict, not once they occur.

Many of the requirements may be met from open-source material or overt means. Examples are sociological and related information about the people, and details about the information infrastructures themselves, especially the NII and GII. Covert collection requirements will also apply, covering aspects of all components (hardware, software, and details about the information, people and power supply) across both the NII and particularly the DII.

High-quality analysis of the intelligence is also critical. This necessity applies across the board, but particularly in psyops related areas where judgements about the people in our region and their responses in various wartime situations and post-war aspirations may be a difficult and, at times, controversial call.

'Wargaming' that identifies intelligence gaps and challenges assessments (as well as developing skills and experience) will form an important part of the intelligence process.

Conclusion

The information society has seen the introduction of rapid technological change that has conditioned how we now live, think, decide and act. From a national perspective, harnessing this change to our national advantage is important, as is protecting our interests from its exploitation by others.

Defence, as for other key national security and wellbeing issues, is particularly important. Defence has acknowledged and embraced the offerings of the new technologies, and is now moving down the path of a networked force that will deliver efficiencies across the whole C4ISREW spectrum. This is also the case in countries within our region, and in other areas where the ADF could be operationally deployed.

In war, our objective would be to win by destroying an enemy's capability to fight as well as their will to fight. This objective entails political, strategic, military, economic, and societal elements and targets. The objectives are both psychological and physical.

A country's NII, DII and GII linkages enable the information society, including its Defence capability. The ability to target and attack the information infrastructures of an enemy in war, or to exploit those of allies, friends or neutrals, must be part of a Defence capability and a national capability. The requirement may also exist to target these infrastructures in non-war circumstances. Developing the capability to target foreign information infrastructures is not only a necessity, but a national priority.

ENDNOTES

[1] An article entitled 'Half the world has a mobile phone', published in the January–February 2008 edition of *ITU News*, the newsletter of the International Telecommunications Union (ITU), stated that the number of mobile phone users was expected to reach 3.3 billion subscribers, or 50 per cent of the global population, in early 2008. This compared with mobile phone ownership by only 12 per cent of the global population in 2000. Developing countries were identified as rising the fastest, with Brazil, Russia, India and China accounting for more than 1 billion subscribers in 2007.

[2] Colonel John Boyd, a US military strategist, devised a four-phase interactive process, known as the *OODA loop*, to simplistically describe the decision cycle. The four phases are: *observe* (what is seen, by whatever means), *orient* (analysis, assessment, knowledge), *decide* (the decision, based on knowledge options), and *act* (action taken or attempted). In addition to its widespread use in a military context, it has also been applied widely in non-military contexts.

[3] Information Operations are described in Australian Defence Doctrine Publication (ADDP) 3-13—*Information Operations* (2006) Glossary of Terms as 'IO is the coordination of information effects to influence the decision-making and actions of a target audience and to protect and enhance our decision-making and actions in support of national interests'. By definition, IO has offensive and defensive aspects. The core purpose of *offensive IO* is to 'influence the decision-making and actions of a target audience'. All or some elements of offensive IO might apply in war or non-war circumstances, and targets include adversaries, neutrals or friends and allies. Targets also could be high-level political/strategic in character, and/or in support of military operations at the operational/tactical level.

[4] ADDP 3-13—*Information Operations*, Glossary of Terms, Australian Defence Headquarters, 2006.

[5] See Defence Signals Directorate (DSD—Australia), *Australia's National Information Infrastructure: Threats and Vulnerabilities*, February 1997. An unclassified version of this report is at Attachment A to a report to the Australian Government by the Attorney-General's Department dated December 1998 entitled 'Protecting Australia's National Information Infrastructure'. This report is available at <http://law.gov.au/publications/niirpt.ptl>.

[6] US Department of Defense, *Department of Defense Dictionary of Military and Associated Terms*, Joint Publication 1-02, 17 October 2007, available at <http://www.dtic.mil/doctrine/jel/doddict/data/g/02329.html>, accessed 28 February 2008.

[7] US Department of Defense, *Department of Defense Dictionary of Military and Associated Terms*.

[8] ADDP 3-13—*Information Operations*, Glossary of Terms.

[9] The *Blackberry* blackout in the United States and Canada in April 2007, lasting about 10 hours and due to a primary server fault, caused considerable confusion and disruption to many thousands of business and government users, especially those who did not have ready access to alternative communications during that period. According to press reports of the incident, many businesses lost considerable amounts of money because of their inability to close deals or exploit market opportunities within critical timeframes.

[10] The concept of modern effects-based operations has largely been developed by Dr Edward A. Smith, Executive Strategist of Effects-Based Operations at the Boeing Corporation. His most recent book on this subject entitled *Complexity, Networking & Effects-Based Approaches to Operations*, Command and Control Research Program (CCRP), Department of Defense, July 2006, is available at <http://www.dodccrp.org/files/Smith_Complexity.pdf>, accessed 26 February 2008.

[11] In today's global marketplace, a critical electronic system might be designed in the United States, comprise operating software written in India, and include physical components manufactured in such countries as China, Malaysia or South Korea.

[12] Identifying who is a threat, knowing their aims and modus operandi, and thus likely targets, is a fundamental part of any associated protective risk assessment and risk management process.

[13] For further details about covert intelligence techniques, see Ian Dudgeon, 'Intelligence Support to the Development and Implementation of Foreign Policies and Strategies', *Security Challenges*, vol. 2, no. 2, July 2006, pp. 61–80, available at <http://securitychallenges.org.au/SC%20Vol%202%20No%202/vol%202%20no%202%20Dudgeon.pdf>, accessed 26 February 2008.

[14] CNE refers to software hacking operations aimed at extracting intelligence from databases within a computer network. CNE is normally a covert activity, and undertaken in such a way as to avoid detection for at least the duration of the intelligence requirements it serves. It differs from Computer Network Attack (CNA), which is hacking into software systems for the purpose of destruction, disruption or degradation of the software, information and/or hardware itself. Both are capabilities under the general heading of Computer Network Operations.

[15] The release on the Internet of a video of the execution and decapitation in February 2002 of Daniel Pearl, the *Wall Street Journal*'s Indian-based South Asia bureau chief, is one example of this. Pearl, a US citizen and Jew, was pursuing a terrorist-related story when he was kidnapped in Pakistan in January 2002. His kidnappers, a group calling itself The National Movement for the Restoration of Pakistani Sovereignty, claimed he was a spy for both the United States and Israel. Western analysts believe one of the reasons for Pearl's assassination was to motivate and recruit new members to the Islamic jihad cause.

[16] See Tamara Makarenko, 'The Crime-Terror Continuum: Tracing the Interplay Between Transnational Organised Crime and Terrorism', *Global Crime*, vol. 6, no.1, February 2004, pp. 129–145, available at <http://www.silkroadstudies.org/new/docs/publications/Makarenko_GlobalCrime.pdf>, accessed 26 February 2008. I am indebted to Professor Peter Grabosky for referring this reference to me. According to Makarenko, many terrorist organisations engage in illegal activity to raise finances. These include the al-Qaeda financial network in Europe, which was reported to be primarily dependent on credit card fraud for funding. However, the primary source of funding for many terrorist groups is the illegal drug trade. Makarenko also claims that since the collapse of the Soviet Union there has been a convergence of transnational organised crime and international networked terrorist groups. This has resulted in the cooperation between terrorist and crime groups where their interests intersect. One example cited was terrorist organisations using criminal organisations to transport and market illegal drugs grown and produced in areas under terrorist control. Another example was the acquisition of guns by criminals on behalf of terrorists.

[17] Alan Sipress, 'An Indonesian Prison Memoir Takes Holy War Into Cyberspace', *Washington Post*, 14 December 2004, p. A19.

[18] For a comprehensive overview of cyber-crime, see Peter Grabosky, *Electronic Crime*, Prentice Hall, Upper Saddle River, NJ, 2007.

[19] The word 'allegedly' has been used as financial institutions generally seek to avoid any publicity about the frequency and type of computer crime, and especially the extent of any major crime, because of the adverse effect it can have on investor or customer confidence. However, one example of a major case of proven fraud was the loss of $A1.6 billion by Barings Bank in Singapore in 1995 by Nick Leeson, their head trader. Leeson conducted rogue trading by falsifying accounts and various misrepresentations that were not detected by internal controls and audit systems.

[20] Sophisticated hacking software is readily available on the Internet, and has been for many years.

[21] An interesting profile of malware writers was in an article by Clive Thompson entitled 'The Virus Underground' published in the *New York Times* on 8 February 2004, available at <http://engineering.dartmouth.edu/courses/engs004/ virusarticle.html>, accessed 26 February 2008.

[22] Colonel John Boyd, *OODA loop* process (refer note 2 above).

[23] A vivid example was the severe anti-US publicity generated in 2004 in the world media in response to reports and photos of the abuse and torture of Iraqi prisoners by some US military police in the Abu Ghraib prison in Baghdad. Anti US and anti-war elements quickly exploited this 'windfall gain', which also gave rise to significant disinformation about other alleged adverse behaviour by US and Coalition partners in Iraq, and US and Western attitudes towards Islam generally.

[24] Dudgeon, 'Intelligence Support to the Development and Implementation of Foreign Policies and Strategies', *Security Challenges*, pp. 61–80.

Chapter 5

Protecting Information Infrastructures

Gary Waters

Introduction

As discussed in chapter 2, the concept of Network Centric Warfare (NCW) anticipates ready access to information. This demands an ability to protect information such that its security can be as assured as its ready access. From a military perspective, therefore, the Australian Defence Force (ADF) must balance the quest for information superiority against the potential for creating an operational vulnerability. And it must do this within the broader context of balancing security and privacy as it increasingly shares information. From a national perspective, the Australian Government will wish to balance these same issues as it shares information across national security agencies and possibly with Non-Governmental Organisations (NGOs).

Reliance on information and information systems and addressing the consequent vulnerabilities raises the question of protecting along a spectrum—from simple failure to attacks from terrorists or state-based adversaries. Cyber-security demands far more attention, and while protecting critical information infrastructure addresses much of the problem, more needs to be done. Government and Defence need to develop a trusted information infrastructure and put in place the mechanism for Australia to protect its critical information infrastructure.

In addressing these issues, this chapter provides frameworks for dealing with operational vulnerabilities, cyber-terrorism, privacy, and security risk, which can be managed at the Defence enterprise level for the ADF and at the whole-of-government level for national issues.

Balancing information superiority and operational vulnerability

The pressure to ensure both a superior information position as well as security of that information within an environment of increased sharing of information has been building with the rise of NCW, increases in coalitions to deal with security threats and challenges, and the focus on domestic security since the 11 September 2001 terrorist attacks on the United States. Information superiority for the ADF will come from:[1]

- seamless machine-to-machine integration of all manned and unmanned systems, including those in space, across the joint force;
- real-time pictures of the battlefield;
- predictive battlespace awareness—driven by commanders who will be required to predict and pre-empt adversary actions when and where they choose;
- assured use of information through effective Information Assurance (IA) and defensive Information Operations (IOs); and
- denial of effective use of information to adversaries through offensive IOs.

Achieving a balance between access to information and protection of that information is something of a contradiction in this age of information and globalisation. Ensuring access to information involves a number of key concepts and assumptions as follows:

- The concepts of 'Information Superiority' and NCW imply that all ADF commanders will have full access to relevant information.
- Commanders at all levels will continue to deal with uncertainty or the 'fog of war' due to a lack of complete and accurate information.
- 'Reach back' will allow support functions to be provided from outside the area of operations, thereby reducing the 'footprint' of the deployed force in the area of operations.
- There will be increasing use of civilian communications for strategic systems into the area of operations and for tactical systems within that area, through out-sourcing and commercialisation. These systems will have to be integrated with any Defence-owned and operated systems and with the systems of other Government agencies.
- The concept of 'national information economy and infrastructure' promotes the efficiencies of computer-based business automation and Internet-based services, which will apply equally to the national security arena.

Improved information accuracy and responsiveness are seen as key attributes that help reduce the probability of failure in modern military operations. The side which is not confounded by the 'fog of war' and which can get inside the decision cycle of the adversary can expect to have a higher probability of success. Information superiority requires connectivity and interoperability across all relevant force elements so that synchronised operations can be carried out at an appropriate operational tempo. The political sensitivities of operations also demand that adequate information and reporting is available at higher levels outside the area of operations.

Vulnerabilities

In terms of vulnerabilities that the ADF could introduce through increased information sharing and access, the following points are relevant:[2]

- Dependence on real-time information and intelligence introduces one form of vulnerability, while the sharing of that information and intelligence introduces another.
- Similarly, dependence on the security and information management policies of coalition partners and other agencies introduces one form of vulnerability, while the new interfaces with those policies of external partners introduces another.
- Key information relating to operations is distributed widely at the strategic, operational and tactical levels and is shared with coalition partners and other agencies.
- Commanders at all levels are exposed to information overload, which can saturate their ability to cope.
- Time-sensitive decisions can be slowed while commanders seek greater clarification of the situation, in the belief that the last piece of information can be found.
- Australian Government networks are dependent on commercial or allied communication systems.
- The adversary, whether State-based or simply an individual, has equal access to modern global communications and the Internet.
- Perceptions generated in political and public minds become reality and, therefore, have to be factored into any strategic and operational planning.
- More information is available at lower levels of command where there is a higher probability of capture by an adversary.

The advantages of access to information through Defence's restricted and secret systems in the normal execution of duties are expected by all personnel. Access to external email and the Internet for daily business is also expected. This dependency introduces potential vulnerabilities associated with the management of a large organisation such as Defence as follows:

- The dependence on email and Intranet for the conduct of business, both inside and outside the Department, means that any disruption of the networks and applications would have an immediate effect on the functioning of Defence.
- External connectivity provides an easy avenue for the manipulation of, and attack on, Defence's information.
- Deliberate or inadvertent misuse of the restricted system for the transmission of classified information outside Defence provides an easy avenue for the media (through whistle blowers, leakers, or paid informants) or adversaries (through sympathisers, agents or supporters).
- Disruption of personnel and pay processes would have an immediate impact on the morale and well-being of military and civilian personnel, more so when they are deployed on operations.

- Some Defence functions are contracted out and depend on e-business arrangements.

Balancing security and privacy in information sharing

Evolving national security policy in both Australia and the United States is seeking higher levels of cooperation, information sharing and system-to-system interaction across Government agencies and between the public and private sectors. Any implementation of trusted information sharing systems will demand that a number of components be addressed:[3]

- collaboratively developed public and private sector policies;
- the use of available frameworks, architectures, standards and technologies; and
- the deployment of systems that integrate effective risk management controls.

All three components are needed in developing trusted systems, which meet three key requirements for privacy and security—government and business requirements as well as citizen expectations.

Where possible, existing bodies of knowledge, current technologies and standards-based products should be used to build and manage the infrastructure for information sharing. In terms of information security, threat management, identity management, access management, and security command and control, capabilities exist today and can be applied effectively once appropriate policies and business processes have been established.[4]

As business privacy and personal privacy continue to become major issues for critical infrastructure protection across both the public and private sectors, an architecture will be needed to address privacy protection and to develop requirements for appropriate information privacy controls. Technologies, such as business intelligence, data management, enterprise management, and storage management systems are currently available to meet requirements.[5] However, more needs to be done to build on the work in the private sector to foster business and personal privacy system architectures and to establish standards-based interoperability.

In the United States, the *National Strategy for Homeland Security* (July 2002) and the *National Strategy to Secure Cyberspace* (February 2003) demanded unprecedented levels of cooperation and system-to-system interaction among private sector companies and all levels of government to ensure protection of national critical infrastructure.[6] This applies equally to Australia and the way in which the National Counter-Terrorism Committee addresses national critical infrastructure issues that reach across the Commonwealth, States and Territories. As John Sabo argues, for the United States:

This will involve new and trusted working relationships among organisations which have had little formal interaction on infrastructure protection issues, as well as on issues such as new systems, expanded network interfaces and significant increases in data collections, data flows, and data integration, analysis and dissemination.[7]

This applies equally in Australia, where common information security and infrastructure protection aspects across Government agencies need to be seen more as a national capability issue rather than simply as an operational support issue.

Managing security risk

There are four key areas in managing security risk—threat management, identity management, access management, and security command and control. These are discussed below.[8]

Threat Management controls are needed to protect networks and systems against external and internal threats. They must also be able to assess vulnerabilities, and identify and mitigate physical and systems-based risks and attacks. Essentially, it is these controls that protect the critical infrastructure.

Identity Management controls are needed to provide a foundation for 'registering' users and for establishing role-based access, as well as enabling role-based and portal views of information and applications. However, uniform policy development will be necessary to address the diverse perspectives of the different organisations involved.

Access Management controls are needed to protect classified, regulated and business-sensitive resources; control how resources are accessed and used; and ensure authorised availability across networks, systems and platforms.

Tiered controls for access management will probably be needed to reflect roles, information classification requirements, and particular organisational rules for information and for participants across the breadth of participating organisations and users. This will become increasingly important if new classification categories for sensitive critical infrastructure information are instigated that fall outside the scope of current security classification levels (such as Secret and Top Secret for Defence and Protected and Highly Protected for non-Defence agencies).

Security Command and Control capability is needed to effectively manage the security of the networked infrastructures. This includes such things as resource management, impact correlation, secure collaboration, intelligent visualisation and predictive analysis tools.

Managing privacy risk

Privacy is defined as 'the proper handling of personally identifiable or business confidential information in accordance with policies having the consent of the data subject or as required by law or regulation'.[9]

Open standards-based architectures, protocols, languages or schemas do not yet exist for ensuring that privacy rules and policies can be embodied in Information Technology (IT) systems or for allowing them to be interoperable across networks that manage the collection and processing of information. Both personal privacy and business privacy requirements must be engineered into the new cyber-security architecture if Australia and the ADF are to develop and field trusted systems.

The vast quantities of data and information that may ultimately flow across network and jurisdictional boundaries will eventually demand automated management of relevant policy rules. These standardised policies and protocols will be needed for scalability, efficiency and trust reasons. Any rules will have to be enforced within an overall governance framework, the start point for which will be a privacy architecture.

There are several already defined services and capabilities that will assist in addressing privacy controls for information sharing. These include:[10]

- control and data usage functionality, to ensure that policies drive business rules processing;
- certification of system credentials;
- validation of data;
- interaction of data subjects, systems and processes;
- individual and business access to data as well as audit capability;
- use of agents, both technology-based and/or human;
- negotiation where appropriate; and
- enforcement of policy violations.

Many of these framework services can be supported by currently available technologies in business intelligence, data management, enterprise management and storage management, particularly when applied to enterprise implementations of data-sharing systems.[11]

Dangers in getting privacy wrong

In March 2004, the Australian Federal Privacy Commissioner, Malcolm Crompton, argued that the case for stronger identity management was being made as a key element for preventing terrorism, preventing identity fraud, and even eliminating spam. Indeed, the case was strengthened as proponents argued how important identity management was for e-business and e-health.[12]

Malcolm Crompton highlighted some of the dangers for the community, business and government of getting privacy wrong when developing identity management solutions:[13]

- Subversion—the natural response of unwilling or suspicious participants, from reduced participation and deliberately misleading information all the way through to more active resistance.
- Pent up reaction that can produce very strong public policy responses, such as the legislative responses in the United States creating the 'Do Not Call' list to keep out direct marketers or the worldwide movement against spam.
- Financial loss when developed projects must be shelved.
- Self-defeating solutions that create new threats to privacy, security and identity integrity.
- Creating the foundation for a total surveillance society, the full implications of which may only be recognised after it is too late.

Technological solutions can be a key part of getting privacy right in this area. Some technologies, such as biometrics, have the potential to enhance privacy depending on how carefully they are designed and implemented. But some of our thinking needs to change, such as the assumption that full identification is needed in all circumstances. Solutions to the issues that Crompton highlights include finding the right answers through technology, law and accountability processes. A critical issue will be the need to fully engage all stakeholders in vigorous public debate along the way.[14]

Search engines, such as Google, have become increasingly more powerful, just as the World Wide Web has become a richer source of information as more individuals, businesses and government agencies rely on it more and more to transmit and share information. The information is stored on servers that are linked to the Internet.

Any errors can lead to this information, which is not meant for public viewing, being made available to the public. Errors can come through improperly-configured servers, inadequacies in computer security systems, or simply human error. It is virtually impossible to pull back the information once a search engine has found it.

An article in the *Sydney Morning Herald* on 18 February 2004[15] highlighted that Google's search engine 'crawled' over every web page on the Internet on a bi-weekly basis. It 'grabbed' not only every page on every public server, but also every link attached to every page, and then catalogued the information.

The article[16] also highlights vulnerabilities which can bring up spreadsheets, credit card numbers and social security numbers linked to a list of customers, as well as total dollar figures in financial spreadsheets. It would be virtually impossible to monitor the tens of millions of searches that occur every day, according to a search engine expert, Tom Wilde.[17] The concern goes even further in terms of the potential for identity theft and identity fraud, as well as other cyber-crime aspects discussed in chapter 3.

Research by AusCERT[18] found that 42 per cent of 200 organisations surveyed had information compromised through attacks on their IT networks. That same research highlighted that in 2001 Optus was breached and 425 000 user names and passwords were compromised.[19]

Cyber-security

Thomas Homer-Dixon postulates a future scenario:[20] In different parts of a US state, half a dozen small groups of men and women gather. Each travels in a rented mini-van to its prearranged destination—for some, a location outside one of the hundreds of electrical substations throughout the state: for others, a point upwind from key, high-voltage transmission lines. The groups unload their equipment from the vans. Those outside the substations put together simple mortars made from materials bought at local hardware stores, while those near the transmission lines use helium to inflate weather balloons with long silvery tails. At a precisely coordinated moment, the homemade mortars are fired, sending showers of aluminium chaff over the substations. The balloons are released and drift into the transmission lines.

Simultaneously, other groups are doing the same thing along the eastern seaboard and in the south and southwest of the United States. A national electrical system already under heavy strain is short-circuited, causing a cascade of power failures across the country. Traffic lights shut off. Water and sewerage systems are disabled. Communications systems break down. The financial system and national economy come to a halt. And if that is not of sufficient concern, Brad Ashley[21] of the US Air Force notes that:

> Today's battlefields transcend national borders. Cyberspace adds an entirely new dimension to military operations, and the ubiquitous dependence on information technology in both the government and commercial sectors increases exponentially the opportunities for adversaries as well as the potential ramification of attacks.[22]

Indeed, Ashley goes well beyond Homer-Dixon's scenario and depicts a number of scenarios all rolled into one devastating attack.[23] Ashley postulates that military systems are under relentless electronic attack and the global media is reporting these attacks with great zeal, thereby adding to the problem. An unknown adversary has seized control of military logistics, transportation and administration systems associated with deployment of forces.

Commercial websites are inundated with requests for connection, which paralyses parts of the Internet. Worldwide computer virus attacks occur, affecting over 60 million computers, including military systems. An orchestrated campaign of individuals flooding Defence and security websites is carried out, a cyber Jihad is started, and national infrastructure computers are infiltrated, leading to raw sewage being released into rivers and coastal waters.

Worse still, Defence networks are penetrated, power grids are infiltrated and shut down, computer problems close the stock market in several capitals. The competitive media helps spread the ensuing panic throughout the world.

If these incidents sound plausible, it is because they have occurred, in varying forms and to varying levels of success over a lengthy period of time. However, were they to be orchestrated over a very short time span as Ashley postulates, their results could be devastating.

To some, the scenarios postulated of Homer-Dixon and Ashley may have sounded far-fetched in 2000; however, the 11 September 2001 terrorist attacks on the United States changed all that. Many nations have since realised that their societies are susceptible to terrorist attacks. There are two trends that explain this: the growing technological capacity of small groups or even individuals to wreak havoc; and the increasing vulnerability of economic and technological systems to quite deliberate and specific attacks.

Adding to the vulnerabilities are the changing communications technologies that now encompass satellite phones and the Internet which permit the coordination of resources and activities across the world. Criminal and terrorist organisations can use the Internet to share information on weapons and tactics, transfer funds, and plan criminal activities or attacks. The links between crime and terrorist organisations mean that any criminal cyber-attack could be financing a terrorist organisation. Identity theft is also cause for concern for banks and financial institutions, as once again a criminal cyber-attack could be linked to a terrorist organisation.

There are several reasons why hackers will seek to gain illegal access to IT systems. These include: to gain financially, to commit sabotage, to steal identities, to commit fraud, to carry out espionage, or to cover up other physical theft. The level of sophistication needed to hack into sites has decreased while the availability of hacking tools has increased substantially. As Ashley notes, adversaries in cyber-space require minimal technology, little training or funding, no infrastructure support, and can launch attacks from anywhere at anytime.[24] A report in 2004 by Trend Micro indicated that viruses affecting personal computers (PCs) cost businesses worldwide some US$13 billion in damages in 2001, US$20 billion in 2002, and US$55 billion in 2003.[25] Add to this, the estimated annual loss due to computer crime of US$67.2 billion, for US organisations alone.[26]

Information-processing technologies have also boosted the power of terrorists by allowing them to hide or encrypt their messages, with the power of a modern lap-top computer today exceeding anything that could have been imagined three to four decades ago. Not only can terrorists and criminals run readily available sophisticated encryption software, they can also use less advanced computer technologies to achieve similar effect. Steganography (hidden writing) that allows people to embed messages into digital photographs or music clips which can then be posted on the World Wide Web for subsequent downloading was reportedly used by terrorists who planned an attack on the US embassy in Paris in 2004.[27]

The World Wide Web also provides ample access to information about critical infrastructure. For example, the floor plans and design of the World Trade Center in New York were readily available, as was information on how to collapse large buildings. Instructions for making bombs and other destructive materials are also readily available. Indeed, practically anything needed on kidnapping, bomb-making, and assassination is now available on-line.[28]

Australia's economic and technological systems make the nation, the Government and the ADF all the more vulnerable because of the interconnectedness across modern society and the increasing geographic concentration of wealth, people, knowledge, and communication links such as highways, rail lines, electrical grids, and fibre-optic cables. As societies modernise, their networks become more interconnected, which means that the number of nodes increases, the links among the nodes increases, and the speed at which things move across these links increases. All of this adds to the rich array of potential targets.

Not only does vulnerability increase through greater numbers, but also the features of interconnected networks can make their behaviour unstable and unpredictable. One obvious example is that of a stock market crash, in which selling drives down prices, which, in turn, leads to more selling. The tight coupling of networks also makes it more likely that problems with one node can spread to others. The United States has experienced a number of cascading effects when electrical, telephone, and air traffic systems have suffered partial failure, which has spread across the country. In addition, the nature of these networks also sees a small shock producing a disproportionately large disruption.[29]

A special commission set up by President Bill Clinton in 1997 reported that 'growing complexity and interdependence, especially in the energy and communications infrastructures, create an increased possibility that a rather minor and routine disturbance can cascade into a regional outage'. The commission continued: 'We are convinced that our vulnerabilities are increasing steadily, that the means to exploit those weaknesses are readily available and that the costs [of launching an attack] continue to drop'.[30]

So much for physical networks: what about psychological networks? Australian citizens are nodes in this network, linked through the Internet, satellites, fibre-optic cables, radio, and television news. Immediately after a crisis, the media and others report the story across this network. Televisions stay on, telephone lines and e-mail messages are used constantly, to the extent that services, especially the Internet, become noticeably slower immediately after the event.

The Australian Government should expect terrorists of the future to target the critical networks that underpin society. This would include networks for producing and distributing energy, information, water, and food; the highways, railways, and airports that make up the nation's transportation grid; and the health care system.[31] While an attack on the food system would be of greatest concern to people, vulnerability of the energy and information networks attract a lot of attention because they so clearly underpin the vitality of modern economies.[32]

The use of Supervisory Control and Data Acquisition (SCADA) systems that monitor and direct equipment at unmanned facilities from a central point pose a worrying potential vulnerability. In 1998, a 12-year old hacker gained control of the SCADA systems that run the Roosevelt Dam in Arizona and, in 2001, a disgruntled worker, Vitek Boden,[33] released waste water in Maroochy Shire, Queensland. More than three million SCADA devices exist throughout the world.[34]

The real concern is that these SCADA networks sit 'squarely at the intersection of the digital and physical worlds. They're vulnerable, they're unpatchable, and they're connected to the Internet'.[35]

SCADA systems are used to digitise and automate tasks such as opening and closing valves in pipes and circuit breakers, monitoring temperatures and pressures, and managing machinery on the assembly line. As these systems connect to corporate networks and as those corporate networks connect to the Internet or adopt wireless technology, the vulnerabilities become more pronounced. The power grid could be taken down, emergency telephone systems could be rendered useless, floodgates to a dam could be disabled, and so on.

These control systems have been designed and developed with efficiency and reliability in mind, not security. Many of the legacy control systems cannot accommodate the newer security technologies such as encryption. Compounding these technical difficulties is a range of cultural and management issues, firmly rooted in the physical world, that pays scant attention to cyber-security concerns.

Initially, SCADA systems were developed with proprietary technology, with no connectivity to corporate networks. However, the impact of globalisation and the Information Age demanded greater efficiency, greater transparency and greater connectivity, which resulted in linking the control networks to corporate networks. This means that hackers who seek to insert worms and viruses in corporate networks can get an additional dividend in that any connectivity to control systems that are not turned off can be affected by the worm or virus.

It was in this way that the Sasser virus disabled several oil platforms in the Gulf of Mexico for two days in 2004, while the SoBig virus affected the rail signalling and dispatching systems of CSX Transportation in August 2003, stopping train services for up to six hours.[36]

While Distributed Control Systems were the predominant form of control systems decades ago, whereby they existed within a small geographic area (say a single manufacturing plant), had all components (hardware, software, master controllers, workstations, etc) provided by the same vendor, and operated over a dedicated Local Area Network, that is no longer the case. The proliferation of SCADA systems across a wide geographic area to distribute oil and electricity in the main sees a lot of master systems communicating with remote devices over the Internet, wireless radio, the public telephone system, or private microwave and fibre-optic networks. The remote units are not only controlled by their master, they also send real-time data back.

The SCADA networks themselves are also vulnerable because of their dependency on the telecommunications that support them. Transmissions could be intercepted and altered, redirected or even destroyed, so the transport medium introduces another area of vulnerability. The use of dial-up modems, where little or no authentication is required, introduces yet another form of vulnerability. Not many companies would operate today without firewalls and Intrusion Detection Systems (IDS) on their IT networks, yet very few have such security mechanisms on their control networks. Even if firewall filters were fitted to the control networks, most firewalls have been designed to filter Internet Protocols (IPs) but not control system protocols.

It is not just about improving SCADA systems, however. More can be done to improve the information security on the corporate networks. Improved router configuration, antivirus software, IDS, and more diligent software patching would all help reduce the vulnerability. There are also non-technology actions that can be taken, such as improved configuration management, better documentation of network architectures, and better contingency planning.[37]

Returning to the broader issue of cyber-terrorism, it is worth noting the US House Armed Services Committee's Sub-committee on Terrorism, Unconventional Threats and Capabilities consideration of 'Cyber Terrorism: The New Asymmetric Threat'[38] on 24 July 2003. The Committee chairman, Jim Saxton, argued that the rapid flow of information was becoming increasingly important on the battlefield. He said that in the nineteenth century three words per minute could be transferred whilst 38 830 soldiers were needed to provide information over

10 square kilometres. In the 1990–91 Gulf War, the transmission rate was increased to 192 000 words per minute whilst only 24 soldiers were needed to cover 10 square kilometres. It is expected that by 2010 the data transfer rate will be further increased to one trillion words per minute whilst only three soldiers will be needed to cover 10 square kilometres.[39]

At the same hearing, Dr Eugene Spafford[40] said that threats from malicious software (malware) had grown steadily for 15 years and threatened military, government, industry, academic and general public information systems. The interconnections across these segments of the community meant that a threat to one could readily spread to the others. His concern is exacerbated by the malware's use of victim computers to carry out the attack, which presents an asymmetric threat to computer systems.

Spafford went on to say that the malware threat to US systems, and the military in particular, is significant because software is at the heart of most advanced systems, spanning weapons, command and control, communications, mission planning, and platform guidance. Furthermore, intelligence, surveillance, and logistics all depend on massive computational resources.[41]

There is also the threat from simple failure that must be factored in. Systems are becoming more complex and much of the software is commercial off-the-shelf (COTS) and not developed to contend with active attacks and degraded environments. Moreover, software vendors have tended to concentrate more on time-to-market as the most important criterion for success, rather than well-designed and well-tested code.[42] Increased connectivity, whereby systems are configured so that every machine has network access, which is needed to provide for remote backups, access to patches, and user access to World Wide Web browsing and e-mail, adds to the threat.[43] Spafford went on to offer a number of recommendations:[44]

- Explicitly seek to create heterogeneous environments so that common avenues of attack are not present.
- Develop different architectures.
- Rethink the use of COTS software in mission-critical circumstances.
- Rethink the need to have all systems connected to the network.
- Require greater efforts to educate personnel on the dangers of using unauthorised code, or of changing the settings on the computers they use.
- Revisit laws that criminalise technology instead of behaviour.
- Provide increased support to law enforcement for tools to track malware, and to support the investigation and prosecution of those who write malware and attack systems.
- Do not be fooled by the 'open source is more secure' advocates. The reliability of software does not depend on whether the source is open or proprietary.
- Initiate research into the development of metrics for security and risk.

- Establish research into methods of better, more affordable software engineering, and how to build reliable systems from components that are not trusted.
- Emphasise the need for a systems-level view of information security. Assuring individual components does little to assure overall implementation and use.
- Establish better incentives for security.
- Increase the priority and funding for basic scientific research into issues of security and protection of software. Too much money is being spent on upgrading patches and not enough is being spent on fundamental research by qualified personnel.
- Most importantly, re-examine the issue of the insider threat to mission critical systems.

There are clearly deficiencies in US and Australian cyber-defences. Malicious and incorrect software pose particular threats because of their asymmetric potential—small operators can initiate large and devastating attacks. The situation cannot be remedied simply by continuing to spend more on newer models of the same systems that are currently deficient. It will require vision and willingness to make hard choices to equip the military and other national security agencies with the defensible IT systems they deserve.[45]

Mr Robert Lentz, Director, Information Assurance, Department of Defense also gave testimony at the hearing,[46] where he argued that a new era of warfare had emerged, through the greater power, agility, and speed afforded by connectivity. Thus, a smaller force can mass combat effects virtually anywhere, anytime through these multiple connections. However, this increasing dependence on information networks creates new vulnerabilities, as adversaries develop new ways of attacking and disrupting friendly forces.

Lentz also described the goals that then Defense Secretary Rumsfeld established for networks, namely to[47]

- develop a ubiquitous network environment;
- richly populate the network environment with information of value, as determined by the consumer; and
- ensure the network is highly available, secure and reliable.

Through these goals, Secretary Rumsfeld was seeking to establish the Department's IA Program—the strategy, policy and resources required to create a trusted, reliable network. While the challenges for IA are substantial because of the size and diversity of the Defence and national security IA community and because IA is both pervasive and interdependent upon many other policies and processes, there are clear opportunities. In the first instance, the policy formulation process could be more open, more visible, more collaborative, and, as a consequence, faster.[48]

Lentz also made the telling comment that the US Administration did not expect to achieve guaranteed protection of its information, systems and networks. However, it had put in place 'a robust Computer Network Defence capability within the Department, a capability that continues to evolve and transform itself in pace with the evolving and transforming threat'.[49]

Finally, Lentz offered a telling reason for factoring legacy systems into strategic planning, by saying that all systems are legacy systems as soon as they go on-line. The demand for greater bandwidth, functionality, connectivity and other features is constantly expanding. Lentz argued that the demand would be met, but that the greater task was to ensure it was met securely. To that end, development of protective technologies for space-based laser, advanced fibre-optic, and wireless transport networks were being pursued, as was the development of end-to-end IA architectures and technologies.[50]

The rate of adoption of Internet-based technology, including dependence on the Internet for voice communications and data distribution, means that nations today have the ability to conduct cyber-warfare.[51] Thus, organisations need to have a strategy for keeping their businesses running, if information systems and facilities that depend on those information systems are unable to operate.

The increasing use of IP networking technology to connect critical infrastructure and the movement to packet-switched voice communications (away from a circuit-based architecture) has increased the vulnerability. Additionally, Voice over Internet Protocol (VOIP) equipment is susceptible to traditional Internet threats like worms, viruses and break-ins from hackers. Denial of Service (DS) attacks, which have been experienced in recent times and taken down websites, could be used to disrupt the flow of voice-carrying packets on an IP network, thereby causing a major breakdown in communications. At the infrastructure level, interfaces that allow maintenance and control of equipment have traditionally been accessed through dial-up modems, and are increasingly being converted to IP network connections.

The Gartner Report[52] identified potential targets as the network interfaces found in equipment used by dams, railroads, electrical grids and power generation facilities, and the interface points between the public switched telephone network and IP networks. Connecting computer systems in banking and finance, law enforcement, rail transportation, and in industries such as chemical, oil and gas, and electrical to IP networking adds to the increasing vulnerability of critical infrastructure.

> Most security technology, when used in conjunction with 'best practices', is appropriate to the proportional risk presented by the threat of cyberwarfare. ... The proportional-risk assumption does not mean that

a cyberwarfare attack would be unsuccessful if undertaken by a determined foe, but that risk is low.[53]

The phrase 'digital Pearl Harbor' has been around since 1995, according to Jim Lewis in 2003, then with the Center for Strategic and International Studies and a former Clinton Administration technology policy official.[54] Lewis considered the threat from cyber-terrorists to have been over-stated. Indeed, work carried out by Gartner in 2003 highlighted that disgruntled insiders, not foreign terrorists, posed the greatest cyber-security threat to companies.[55]

Even the most comprehensive IT security technology cannot stop the careless, uninformed, or disgruntled person with access to the network from wreaking havoc. 'The fact is that some of the most devastating threats to computer security have come from individuals who were deemed trusted insiders'.[56]

Costs associated with security policies and software are significant enough, without having their effect decreased by insiders who may not fully appreciate their role in maintaining a secure enterprise. The main reasons behind internal security breaches are noted as ignorance, carelessness, disregard for security policies, and maliciousness.[57] Hence, the best way to address the potential for such breaches is through an awareness and education program, aimed at reducing the effect of 'social engineering'.

Social engineering plays upon the inherent trust that people have in one another and their basic desire to help others. Social engineering tactics will not work if people are informed and aware. Thus, employees should not open unsolicited email attachments and they should scan attached documents for a virus before opening them. They should be aware that attackers will seek to take advantage of a natural trust in sharing files. Employees who use Internet Relay Chat and Instant Messaging services should know about ploys that might be used to lure them into downloading and executing malware that would allow an intruder to use the systems as attack platforms for launching distributed DS attacks. Employees should treat with extreme caution any requests for passwords or any other sensitive information.[58]

Richard Hunter of Gartner has cautioned companies to alert their employees against social engineering. Hunter's view is that the most successful ways for foreigners to steal US secrets is to use such practices or to buy US companies in possession of secrets. After all, computer hacking constitutes only 6 per cent of theft attempts.[59]

At a conference in 2004, concern was expressed over US federal agencies not securing their computer networks and failing to factor technology security into long-term planning. House Government Reform Committee Chairman, Tom Davis, called for increased investment in IT security infrastructure, but acknowledged that the appropriations process 'is always about the here and now'.[60] The problem is, of course, that information network defence requires long-term investment and top-level attention, which is not a natural by-product of the annual budgetary cycle.

The Internet continues to hold so much promise, but, according to the *Economist*, it has to become more trustworthy if it is to realise its full potential.[61] Detracting from trust in the Internet is the continuing worm and virus attacks such as the Blaster worm and SoBig virus that attacked in 2003, causing estimated losses of US$35 billion.[62] As the uptake of broadband increases and as more PCs and other devices are connected, the potential fall-out from further virus, or the more insidious worm, attacks can only increase.

The speed with which these attacks can be launched is also increasing (i.e. attacks are happening faster). The time from initial disclosure of a flaw to the attack by the Slammer worm in January 2003 was six months, which halved the time taken in the previous year. For the Blaster worm in August 2003, the time had fallen drastically to three weeks.[63] Over 500 000 computers were infected and CSX Corporation had to stop its train services as its rail signalling system was brought down, and check-in services of a number of major airlines were disrupted.[64]

Worse still, the intensity of attacks has increased, with the Slammer worm infecting 90 per cent of vulnerable computers within 10 minutes.[65] The network-security monitoring firm, Qualys, has argued that most organisations take on average one month to patch their known vulnerabilities, whereas future attacks could inflict their intended damage within a couple of minutes.[66]

On 27 January 2004, the world experienced the MyDoom virus (also known as Norvarg or Shimgapi). It was immediately rated as a high-level security threat, geared as it was around mounting DS attacks on SCO's website (a US software company). Attacks, such as this, which aim to bring down a company's systems by flooding them with traffic, could very well be precursors to cyber-attacks by nations or terrorist organisations.

Indeed, John Donovan's (Managing Director of Symantec—an Internet security company) research indicated that politically motivated attacks were likely to increase.[67] The attack on SCO was even more insidious as MyDoom left a communications port open on the infected computer, which could have been remotely accessed by a hacker.

Furthermore, as Robert Lemos (a staff writer for CNET News.com) argued, such a virus allowed hackers to hide their real locations, thus making it very difficult to trace any on-line attack. The Code Red virus infected many computers in July 2001, with tens of thousands still infected in 2004 (according to Lemos).[68]

The Sobig.F virus of August 2003 accounted for one out of every 17 email messages and infected over 570 000 computers, while MyDoom accounted for one in 12. Message Labs (a company that filters email for corporate customers) had detected and quarantined more than 1.5 million infected emails within 27 hours.[69] The Sobig virus could have launched an Internet-wide attack had its programming been so designed.[70]

The dramatic increase in cyber-incidents can be seen from the following statistics—between 1995 and 2005, the reports to Carnegie Mellon's Computer Emergency Response Team increased from 171 incidents to 5990.[71]

Trust in the Internet is also undermined through fraud and spam. Indeed, the statistics quoted by the *Economist* are alarming—citing that some 10 per cent of all emails were scams of one sort or another.[72] The degree of cunning in much of this fraud is worrying; for example, brand spoofs that claim to come from trusted companies, fake web pages, fake press releases, and 'phishing'—tricking recipients into giving out sensitive information, such as credit-card numbers, pin numbers and passwords.

Most companies, government agencies and indeed a number of private individuals are now using firewalls to keep malicious code out of their internal networks, and IDS that analyse what gets past the firewalls. Anti-virus software has become commonplace, although there remains a concern over how up-to-date that software is.[73]

While many argue that greater government intervention is needed, that is likely to simply drive up the cost of being connected. Others argue that software vendors should be liable for its security—in other words, vendors should be writing simpler, safer software. So, perhaps, the solution is a combination of both, whereby government legislates that vendors are liable. This would then compel software companies to carry product-liability insurance. Insurance companies would respond by pricing the risk, whereby software companies that write safer code would have an economic advantage.[74]

Another option might be to eliminate Internet anonymity, such that every user could be traced.[75] One way of doing this might be to authenticate each email before it can be sent, by referring to a driving licence, passport, tax file number, social-security number, or some other trusted form of identification.[76]

As Ed Waltz observes, by using a basic risk management approach we can aim to prevent access to 80 per cent of possible attacks.[77] We can detect the presence of the remaining 20 per cent, noting that we would seek to contain 19 per cent of those attacks, and aim to have in place the recovery mechanisms for the 1 per cent that are not prevented, detected or contained.[78] Even with this methodology in place, we must acknowledge that there may be attacks from which we cannot recover and, therefore, we also need to cater for that residual of less than 1 per cent.[79]

Functions that are needed to support protection include monitoring the information infrastructure; generating alerts if an attack is detected or anticipated; controlling the response to modify protection levels or restore service if an attack has been carried out; conducting forensic analysis (including attack patterns, attacker behaviour, damage, and so forth); and reporting to higher authority.[80]

The potential for individuals, organisations or nation-states to mount an information attack with the intent of exploiting, disrupting, or manipulating Australian Government or ADF operations is increasing, to the extent that some analysts have coined the term 'weapons of mass effect', because they can threaten national interests.[81] Hence, it would be prudent for the Australian Government and the ADF to develop the capabilities for discerning, deterring and defending against such threats.

The Australian Government recognises the country's increased vulnerability to acts of cyber-terrorism and other e-security threats because of the nation's growing dependence on the information economy. Accordingly, the Government has designed an e-security policy framework to[82]

- enhance e-security awareness and practices amongst home users and the business community;
- promote the security of Australia's national information infrastructure through information sharing and collaboration with the private sector;
- ensure the government's electronic systems are appropriately secure; and
- promote the security of the global information economy through international engagement.

The Australian Government has also enacted the *Cybercrime Act 2001* to 'prosecute groups who use the Internet to plan and launch cyber-attacks that could seriously interfere with the functioning of the government, financial sector and industry'.[83] The Government's definition of cyber-attacks includes activities such as hacking, computer virus propagation and DS attacks.

Computer Emergency Response Teams (CERTs) have been set up internationally to improve computer systems' security. Australia has set up a team, AusCERT. This is a not-for-profit body operated by the University of Queensland. The Attorney-General's Department also has the Australian Government Computer Emergency Readiness Team (GovCERT.au) that

> develops and coordinates government policy for computer emergency preparation, preparedness, response, readiness and recovery for major national information infrastructure incidents. It also acts as a point of contact within the Australian Government for foreign governments on CERT issues, and coordinates any foreign government requests.[84]

Australia is also leading an Asia-Pacific Economic Cooperation (APEC) initiative to build CERT capacities in developing economies.

The Australian Federal Police (AFP) hosts the Australian High Tech Crime Centre, which investigates e-security incidents in public and private sector organisations. The Centre 'performs a national coordination role for the law enforcement effort in combating serious, multi-jurisdictional crime involving complex technology'.[85]

While the Australian Government and the Australian business sector have established solid risk management guidelines and adhere to sound international risk management standards, Heinrich de Nysschen argues that:

> in future a concerted effort will have to be maintained, building on current efforts, involving all stakeholders, to develop proactive and reactive IT risk management strategies. Only then could we ensure that Australian IT systems, infrastructure and assets are secure, and able to effectively mitigate the impact of potential future security incidents.[86]

Heinrich de Nysschen's view tends to be echoed by comments in 2006 from the US Cyber Security Industry Alliance, which argued for a short list of high priorities on communications and cyber-security to be addressed very quickly.[87] First, a more aggressive research and development program to build secure information systems is needed to mitigate the risk. Of the US$1 billion science and technology budget for the US Department of Homeland Security (DHS) in 2007, only US$20 million is earmarked for cyber-security.[88] The second priority is an early-warning system, while the third is the ability to assure communications bandwidth in an emergency. The fourth priority is a plan to recover the Internet after a disaster and to cope with the interim.

Critical Infrastructure Protection in Australia

Critical infrastructure covers

> those systems we all rely on in our day-to-day lives—communications networks, banking, energy, water and food supplies, health services, social security and community services, emergency services and transport. These are the physical facilities, supply chains, information technologies and communication networks, which, if destroyed or degraded, would adversely impact on Australia's social or economic wellbeing, or affect our ability to ensure national security.[89]

It also includes key government services and national icons.[90]

The continuity of supply of all critical infrastructure is dependent, to some extent, on availability of other infrastructure. Indeed, some sectors are mutually dependent on one another. The degree and complexity of interdependencies is increasing as Australia becomes more dependent on shared information systems and convergent communication technologies, such as the Internet.

The Australian Government is seeking 'to ensure that there are adequate levels of protective security for national critical infrastructure, minimal single points of failure and rapid, tested recovery arrangements'.[91] Furthermore, the government sees its role as providing strategic leadership on national critical infrastructure protection through the Attorney-General's Department. This department is responsible for

> providing national coordination in areas of joint Commonwealth, state and territory responsibility, producing and communicating relevant information to key government and non-government stakeholders, promoting critical infrastructure protection as a national research priority and leading Australia's international engagement on critical infrastructure protection issues.[92]

Compounding the challenge for government is that in some instances, around 90 per cent of critical infrastructure is privately owned. Individual companies are unlikely to have the information or resources to address the risks from a whole-of-sector perspective. Clearly, critical infrastructure protection can only be carried out by a mix of government at all levels and private companies and their industry affiliations. This is as much an awareness activity as it is a risk assessment and mitigation activity. Indeed, this statistic has led the Government to argue:

> The primary responsibility for the protection of Australia's critical infrastructure rests with infrastructure owners and operators. ... Protecting Australia's critical infrastructure therefore requires high levels of cooperation between business and government at all levels.[93]

The former Australian Prime Minister, John Howard, told a business forum on 23 June 2004 that 'we now live in a world where taking measures to improve security and fight terrorism is a cost of doing business'.[94] The *Australian Financial Review* noted 'concerns in the government' that both the Critical Infrastructure Advisory Council and the Trusted Information Sharing Network (TISN) would be more effective through greater engagement of Chief Executive Officers in counter-terrorism consultations.[95]

The TISN strategy provides an overarching statement of principles for critical infrastructure protection in Australia, outlines the major tasks and assigns the necessary responsibilities across government, the owners and operators of infrastructure, their representative bodies, professional associations, regulators and standards setting institutions.[96] The TISN allows owners and operators of critical infrastructure to share information on issues related to the protection of critical infrastructure within and between their respective industry sectors.

These issues include business continuity, consequence management, information system vulnerabilities and attacks, e-crime and the protection of key sites.

The Australian Government has established the Computer Network Vulnerability Assessment Program, which

> provides co-funding on a dollar-for-dollar basis to help owners and operators of critical infrastructure identify major vulnerabilities within computer systems and dependencies between computer networks, and to test the ability of systems to resist exploitation.[97]

To counter this increase in vulnerability, both the public and private sectors are putting in place stronger authentication and identity management for IT systems. Two-factor authentication is coming into vogue, whereby a password or Personal Identification Number (PIN) is used together with an authenticator such as a secure ID token, smart card, or digital certificate.

The government has also created the Business Government Advisory Group on National Security, which 'provides senior business leaders with an opportunity to discuss the strategic direction of our national security policy and provide advice and feedback on national security issues relating to critical infrastructure protection'.[98] This group is chaired by the Attorney-General.

The 2004–2005 Australian Federal Budget allocated further funding to support the Government's Critical Infrastructure Protection strategy, including the TISN. The 2004 budget funding was designed to assist the telecommunications (including Internet Service Providers (ISPs), broadcasting and postal industries in improving the protection of Australia's communications infrastructure through improved information sharing and cooperation.

Funding in this budget also supported a number of groups[99] such as the Communications Sector Infrastructure Assurance Advisory Group (CSIAAG),[100] the Critical Infrastructure Advisory Council (CIAC) and the National Counter-Terrorism Committee and related groups such as the Information Technology Security Experts Advisory Group.[101] The Australian Broadcasting Authority also received funding to assist in developing and maintaining effective communications mechanisms with broadcast operators for critical infrastructure protection coordination.

There were a number of significant new security-related initiatives in the Australian Government's Budget speech of 9 May 2006. The AFP received ongoing research and development capacity to counter the use of new and emerging technologies by terrorists and it established a single facility to manage the collection, monitoring, recording and evidence preparation of terrorism-related electronic surveillance material. Part of the significant allocation to the Australian Security Intelligence Organisation (ASIO) was designed to improve its IT networks to cope with the new demands and expanded operations.

Preventing identity theft also received attention in the 2006 budget. Both a national Document Verification Service and Identity Security Strike Teams were set up. The new service allows government agencies to check Australian passports, the Health Services access cards, Australian citizenship certificates, birth certificates and drivers' licences issued in Australia.

Funding was also provided for key law enforcement and security agencies to ensure a continued capability to intercept telecommunications; further funding was allocated to improve communications within government and with the public during a national crisis; and funding was provided to establish a National Emergency Call Centre capability that can be operational with one hour's notice of an emergency of national significance being declared by the government.

In 2007, the Australian Government revised its *E-Security National Agenda*,[102] releasing the new version in July that year, and allocating a budget of A\$73.6 million over four years. IT security in critical infrastructure, individuals at home and companies are now considered to be highly interrelated. The government has recognised that poor PC security can lead to home computers being used in Distributed Denial of Service (DDS) attacks on critical infrastructure and government agencies. The increase in the sophistication of e-security attacks has made it more difficult for anti-malware software companies to identify attacks and protect clients.

The Agenda created a single whole-of-government committee—the E-Security Policy and Coordination (ESPaC) Committee—to replace two former committees run by the Department of Communications, Information Technology and the Arts (DCITA) and the Attorney-General's Department. The Agenda has three priorities:

- reducing the e-security risk to Australia's national critical infrastructure;
- reducing the e-security risk to Australian Government information and communication systems; and
- enhancing the protection of home users and Small-to-Medium Enterprises (SMEs) from electronic attacks and fraud.

In respect of the first priority, the operations of GovCERT.au will be expanded significantly through the addition of 10 more staff members. It will also have responsibility for the Computer Network Vulnerability Assessment Program, which supports critical infrastructure owners and operators in checking network security.

Government will consider establishing a dedicated Centre to share security information between government and critical infrastructure organisations so as to minimise the impact of electronic attacks.

The AFP will expand its activities in combating on-line criminal activity, including enhancing its ability to detect, deter and investigate criminal threats against critical infrastructure and for technology enabled crime such as on-line fraud.

In respect of the second priority, the Defence Signals Directorate (DSD) will receive increased funding to improve its technical advice on IT security issues for government agencies, managing e-security breaches for agencies, and analysis of malware to rapidly develop countermeasures. The Australian Government Information Management Office (AGIMO) has been commissioned to establish a single framework for the continued delivery of government services in the event of a disruption and/or failure of government-operated information, communication and technology systems.

In respect of the third priority, the Australian Communications and Media Authority (ACMA) will expand its work with Australian ISPs to help them identify compromised computers of their clients. The DCITA will continue to develop and expand its information to home and SME users, delivering information via www.staysmartonline.gov.au.

Notwithstanding these significant initiatives, more effort needs to be put into developing a reliable national indicators and warning architecture, and to improve national planning, programming and operations to build the capabilities needed to discern, deter and defend against the spectrum of cyber-threats that loom on the national security horizon. Changes in the nature of computers and networking could improve processing power, information storage, and bandwidth to the extent that artificial intelligence could be applied to cyber-warfare.[103]

The specific requirements for developing a national indicators and warning architecture for infrastructure protection would be in terms of facilitating the following:[104]

- an understanding of baseline infrastructure operations;
- the identification of indicators and precursors to an attack; and
- a surge capacity for detecting and analysing patterns of potential attacks.

Greater analytical expertise is needed within Defence and other government agencies to address the challenges related to information-based attacks. Rapid attribution of cyber-events is critical to mitigating attacks and deterring future ones. This means mature forensic capabilities are needed to support the attribution and the necessary legal regime to allow for rapid apprehension and prosecution. International deficiencies such as uniform laws that criminalise cyber-attacks and protocols for enforcing laws also need to be addressed.[105]

Furthermore, all federal departments and agencies should be tasked with developing and submitting plans for protecting the physical and cyber-critical infrastructure and key resources that they own or operate. The plans should address identification, prioritisation, protection, and contingency planning, including the recovery and reconstitution of essential capabilities.[106]

Securing the Defence enterprise

From the discussion thus far, it is obvious that as Defence, like other enterprises, reaches out with its networks and is accessed by ever-growing numbers of friends, partners and adversaries, the risk of misuse, theft or sabotage increases. A suitable framework for addressing the vulnerabilities outlined in this chapter, and for securing the Defence enterprise, might be in terms of four integrated layers of activity—policy, operations, systems, and technical measures.[107]

In policy terms, the ADF would need to address such issues as the design, planning and implementation of communications and information systems, which is a collaborative activity between users and providers to achieve a negotiated service. Force protection consequences should be more important than information access, which means that information might have to be restricted or even withheld from a user who has a high probability of capture or compromise. Active information governance measures such as responsibility, authority, procedures, contingency arrangements, reporting and standards should apply across the network.

In operations terms, the ADF would need to address connectivity and interoperability associated with joint force operations as well as combined force operations. In the former, all force elements would need to be connected at the lowest practicable organisational levels (e.g. infantry patrol to close air support aircraft). In the latter, connectivity might be between components and selected force elements (e.g. ADF land component commander to US amphibious task group). Finally, connectivity to Defence finance, logistics and personnel systems and to the systems of other agencies is required by deployed forces.

Systems integration and interoperability will minimise duplication and single points of failure. Cryptographic security, security against computer network attack, and personnel and infrastructure security arrangements should all be provided to the lowest level of connectivity. Robust system redundancy should be provided with appropriate levels of survivability and recovery, and preventive security measures should be offered through enhanced deterrence, detection, containment and response services.

In technical terms, a number of possible initiatives present themselves. First, IT systems (including communications and cryptographic) standards, configuration and protocols should be made compatible with national and combined requirements. Second, dynamic system security can be achieved through appropriate cryptographic, firewall, and virus protection, while dynamic system survivability can be achieved through appropriate routing, standby and duplicate equipment and services. Third, coalition IT and communications standards should be compatible with commercial requirements. Fourth, classification, storage, release and distribution arrangements should be made that also include training, processes, procedures and responsibilities.

Other technical matters such as security architectures, secure identities and access, secure workforce, secure content management and secure web services also need to be addressed. These are covered in more detail below.[108]

Integrated security architectures need to cover directory services, Public Key Infrastructure (PKI), and privilege management infrastructure, as well as digital signatures, authentication, access control, network security, workstation and Personal Digital Assistant (PDA) security, application security, and monitoring, IDS and incident response systems.

Identity and access management needs to cover all aspects of authentication, authorisation and entitlement. Access should be granted only for authorised users, and those users should access only that information they need to access.

Increasingly, workers will be more mobile and their access will need to be secured. Similarly, portals and email systems add to overall vulnerability, which in turn demands greater security vigilance. While web services provide real-time integration of business services from multiple sources, they also add even further to network vulnerability.

As part of this framework, Defence also needs a strong risk assessment methodology (covering attack and penetration testing, and emergency response measures), solid infrastructure security (by designing secure networks, perimeter security controls, multi-layered anti-virus architectures, secure wireless networks and remote access points, and system hardening), business continuity and the ability to recover from shocks and disruptions. Just as importantly, any enterprise with which Defence interacts electronically needs to have in place these security features.

Trusted information infrastructure

One way of addressing trusted information infrastructure is to develop a data access and management system that incorporates enterprise security, identity management, IA and information dissemination management. At the technical level, this would mean data standardisation, encryption and PKI tagging, and a protected data fusion engine that could manage the secure authentication process.

As Philip Dean and Bruce Talbot[109] suggest, such a system would provide a secure place to post classified information that would be accessible from networks of various classifications, all within a securely managed workflow that would ensure that trust could be managed, assured and controlled.

COTS software would be sufficient for providing Multi-Level Identity Management and Secure Service Provisioning. These two concepts would need to be developed in tandem to ensure that security could be delivered through a layered approach that also manages identification, security clearance and access rights of both providers and users of the information. Location, information access and physical protection would be afforded by:

- providing a posting area for information that could be fully managed and secured and that could only be accessed by authorised users;
- offering compartmented storage within that posting area as necessary; and
- tagging devices such as PCs to ensure that they meet device constraints related to the specified information.

A data standardisation regime would be needed to ensure data that had been posted could be received by all authorised devices. Additionally, a management standardisation regime would be needed so that all interactions could be managed, such as the posting of documents, the identities of information providers and users, and the flow of information (based on policies, rules and identity).

Just as intelligence, command and control, and corporate information systems cry out for multiple layers of security to improve information sharing and collaboration and to reduce costs, so too do the interoperability requirements within the battlespace. Specifically, mission control systems such as fire-control systems on naval surface combatants and the multi-function displays in combat aircraft will need to be linked to ground forces in future in order to deliver integrated joint fires.[110]

While the foregoing is aimed squarely at Defence, the same issues pertain to a whole-of-nation perspective to securing Government agencies and ensuring a networked trusted information infrastructure.

Addressing the national requirement

Australia is confronted with a dynamic strategic environment that is continuously evolving. In meeting the exigencies of that environment, Australia will rely increasingly on the power it derives from networked government agencies and the processes for whole-of-government approaches to threats and problems. These exigencies will arise with little notice, impacting on national interests at home and overseas, and may originate from home soil as much as from another country.

As Australia finds itself engaged in persistent operations, where traditional distinctions blur between peace and war, combatants and non-combatants, and foreign and domestic activities, it will need to be perpetually reassessing any strategic gaps both in its preparedness to act and its actual performance on the day.[111] Accordingly, Australia needs the ability to focus, shape, and guide national effort across its networks. That national effort can no longer be permitted to be fragmented in its organisation and disjointed in its application.[112] Any national effort must incorporate an offensive dimension as well as a defensive one, as well as preventive and responsive policies.[113]

From a military perspective, we are clearly moving into the fourth generation of warfare, a generation we might term 'net-war'. The first generation involved massed manpower, the second massed firepower, and the third manoeuvre. Net-war will be characterised by antagonists who will fight in the

> political, economic, social and military arenas and communicate their messages through a combination of networks and mass media. This generation is likely to be based more on ideas rather than military technology; this is a crucial point. Warfare will not be the relatively clear-cut, high technology 'stately dance' of conventional war but rather extremely complex, mainly low-intensity conflicts. In these conflicts it will be hard to differentiate between war and peace, military operations and crimes, front and rear areas, combatants and non-combatants. Fighting will involve an amalgam of military tactics from all four generations and the concepts of 'victory' and 'defeat' will probably cease to exist.[114]

When these pressures are combined with where the ADF is moving with respect to NCW, there are compelling reasons to oversee developments from a single organisational perspective. A Net-war or Cyber-warfare Centre would provide just this—ensuring a joined-up national effort that incorporated offensive, defensive, preventive and responsive strategies, policies and actions while supporting the development and protection of robust networks that underpin the ADF's NCW capability. Such a Centre would be responsible for all aspects of operational planning, support and training, as well as research and capability development not only for the ADF but across national security agencies as a whole. Moreover, the Centre would have a key role in supporting Australia as network complexity and national, allied and coalition Internet-working increase in the years ahead.

Conclusion

Information is the lifeblood of the ADF's future networked force and, as such, it must be protected. The quest for information superiority to underpin network-centric operations in future will introduce operational vulnerabilities. The challenge is to identify these vulnerabilities and develop a framework to address them (in policy, operations, systems, and technical terms). Information sharing is crucial for networked, dispersed forces, but as these forces reach back into their enterprise systems and across into others, the issue of privacy looms large and must be managed through sensible architectures.

The security of information and the underpinning technology is compounded by the threat of cyber-attack, which demands a sophisticated protection system for monitoring the information infrastructure, issuing alerts, and controlling responses as necessary. It would seem timely for national planning, programming and operations to build the capabilities needed to discern, deter and defend against the spectrum of cyber-threats that loom on the national security horizon.

Reliance on information does not just introduce vulnerabilities in the ADF and in its enterprise systems; it also introduces vulnerabilities in all critical information infrastructure on which Australia relies. Defence must have in place robust information security mechanisms across its networks, and it must also ensure that similar security mechanisms exist in other enterprises with which it seeks connectivity.

A trusted information infrastructure is key to supporting Information Superiority and Support (IS&S), which in turn, is key to a networked ADF. That Trusted Information Infrastructure involves much more than just the ADF's networks, and must address enterprise security, identity management, IA, and the management of dissemination of the information. And it applies as much to all Australian Government agencies as it does to Defence.

As network complexity and national and coalition Internet-working increase in the years ahead, there will be pressure on the Australian Government to ensure a joined-up national effort that incorporates both offensive and defensive policies and actions through some form of a Cyber-warfare Centre. Defence could take the initiative now and set up such a Centre to support the development and protection of its robust networks needed to underpin its NCW capability.

ENDNOTES

[1] I thank my colleague Brigadier Steve Ayling for his assistance in clarifying some of this thinking in 2004 and 2005.

[2] My thanks to Steve Ayling for his informed discussions on helping me identify these vulnerabilities.

[3] John T. Sabo, *Addressing a Critical Aspect of Homeland Security: Managing Security and Privacy in Information Sharing Systems*, Computer Associates White Paper, January 2004, available at <http://www.ehcca.com/ presentations/privacyfutures1/4_01_2.pdf>, accessed 4 April 2008, p. 2.

[4] Sabo, *Addressing a Critical Aspect of Homeland Security: Managing Security and Privacy in Information Sharing Systems*, p. 2.

[5] Sabo, *Addressing a Critical Aspect of Homeland Security: Managing Security and Privacy in Information Sharing Systems*, p. 2.

[6] Sabo, *Addressing a Critical Aspect of Homeland Security: Managing Security and Privacy in Information Sharing Systems*, p. 2.

[7] Sabo, *Addressing a Critical Aspect of Homeland Security: Managing Security and Privacy in Information Sharing Systems*, p. 3.

[8] Sabo, *Addressing a Critical Aspect of Homeland Security: Managing Security and Privacy in Information Sharing Systems*, pp. 4–5.

[9] Sabo, *Addressing a Critical Aspect of Homeland Security: Managing Security and Privacy in Information Sharing Systems*, p. 5.

[10] Sabo, *Addressing a Critical Aspect of Homeland Security: Managing Security and Privacy in Information Sharing Systems*, p. 6.

[11] Sabo, *Addressing a Critical Aspect of Homeland Security: Managing Security and Privacy in Information Sharing Systems*, p. 6.

[12] The Federal Privacy Commissioner Malcolm Crompton completed his five-year term at the Commission on 19 April 2004. Just prior to his departure, he delivered a keynote farewell address entitled 'Proof of ID Required? Getting Identity Management Right' in Sydney to the Australian IT Security Forum on 30 March 2004, available at <http://www.privacy.gov.au/news/speeches/sp1_04p.pdf>, accessed 3 March 2008.

[13] Crompton, 'Proof of ID Required? Getting Identity Management Right'.

[14] Crompton, 'Proof of ID Required? Getting Identity Management Right'.

[15] See 'Privacy exposed', *Sydney Morning Herald*, 19 February 2004, available at <http://smh.com.au/articles/2004/02/18/1077072702295.html>, accessed 3 March 2008.

[16] 'Privacy exposed', *Sydney Morning Herald*, 19 February 2004.

[17] 'Privacy exposed', *Sydney Morning Herald*, 19 February 2004.

[18] AusCERT is based at Queensland University and is a non-profit computer security organisation.

[19] 'Privacy exposed', *Sydney Morning Herald*, 19 February 2004.

[20] Thomas Homer-Dixon, 'The Rise of Complex Terrorism', *Foreign Policy*, Issue No. 128, January//February 2002, pp. 52–62.

[21] Colonel Ashley was the Chief of Plans, Policy and Resources Division in the Communications and Information Directorate of Headquarters Pacific Air Forces, Hickham Air Force Base, Hawaii.

[22] Colonel Bradley K. Ashley, US Air Force, 'The United States is Vulnerable to Cyberterrorism', *SIGNAL*, March 2004, p. 61.

[23] Ashley, 'The United States is Vulnerable to Cyberterrorism', *SIGNAL*, p. 61.

[24] Ashley, 'The United States is Vulnerable to Cyberterrorism', *SIGNAL*, pp. 62–63.

[25] Heinrich de Nysschen, 'Homeland Security', *Image & Data Manager*, May/June 2005, p. 36.

[26] US Government Accountability Office (GAO), *CYBERCRIME: Public and Private Entities Face Challenges in Addressing Cyber Threats*, GAO-07-705, Report to Congressional Requesters, Washington, DC, June 2007, available at <http://www.gao.gov/new.items/d07705.pdf>, accessed 4 March 2008.

[27] Homer-Dixon, 'The Rise of Complex Terrorism', p. 2.

[28] Homer-Dixon, 'The Rise of Complex Terrorism', p. 3.

[29] Homer-Dixon, 'The Rise of Complex Terrorism', pp. 3–4.

[30] Homer-Dixon, 'The Rise of Complex Terrorism', p. 4.

[31] Homer-Dixon, 'The Rise of Complex Terrorism', pp. 5–6.

[32] Homer-Dixon, 'The Rise of Complex Terrorism', p. 6.

[33] 'Fighting the worms of mass destruction', *Economist*, 27 November 2003, available at <http://www.economist.com/science/displayStory.cfm?story_id= 2246018> and on the Computer Crime Center website at <http://www.crime-research.org/library/Analitic_nov1.html>, 28 November 2003, accessed 3 March 2008.

[34] Ashley, 'The United States is Vulnerable to Cyberterrorism', *SIGNAL*, p. 64.

[35] Todd Datz, 'Out of Control', *CSO*, vol. 2, no. 1, 2005, p. 28.

[36] Datz, 'Out of Control', *CSO*, p. 30.

[37] Datz, 'Out of Control', *CSO*, p. 32.

[38] Jim Saxton, opening statement before the House Armed Services Committee on Terrorism, Unconventional Threats and Capabilities; hearing on 'Cyber Terrorism: The New Asymmetric Threat', 24 July 2003, available at <http://www.iwar.org.uk/cip/resources/status-of-dod-ia/03-07-24saxton.htm>, accessed 3 March 2008.

[39] Saxton, opening statement at hearing on 'Cyber Terrorism: The New Asymmetric Threat'.

[40] Eugene. H. Spafford, testimony before the House Armed Services Committee on Terrorism, Unconventional Threats and Capabilities; hearing on 'Cyber Terrorism: The New Asymmetric Threat', 24 July 2003, available at <http://www.iwar.org.uk/cip/resources/ status-of-dod-ia/03-07-24spafford.pdf>, accessed 3 March 2008.

[41] Spafford, testimony at hearing on 'Cyber Terrorism: The New Asymmetric Threat'.

[42] This is referred to as competing 'in Internet time'.

[43] Spafford, testimony at hearing on 'Cyber Terrorism: The New Asymmetric Threat'.

[44] Spafford, testimony at hearing on 'Cyber Terrorism: The New Asymmetric Threat'.

[45] Spafford, testimony at hearing on 'Cyber Terrorism: The New Asymmetric Threat'.

[46] Robert F. Lentz, testimony before the House Armed Services Committee on Terrorism, Unconventional Threats and Capabilities; hearing on 'Cyber Terrorism: The New Asymmetric Threat', 24 July 2003, available at <http://www.iwar.org.uk/cip/resources/status-of-dod-ia/03-07-24lentz.htm>, accessed 3 March 2008.

[47] Lentz, testimony at hearing on 'Cyber Terrorism: The New Asymmetric Threat'.

[48] Lentz, testimony at hearing on 'Cyber Terrorism: The New Asymmetric Threat'.

[49] Lentz, testimony at hearing on 'Cyber Terrorism: The New Asymmetric Threat'.

[50] Lentz, testimony at hearing on 'Cyber Terrorism: The New Asymmetric Threat'.

[51] See Antone Gonsalves, 'Gartner: Dependence On Internet Boosts Risks of Cyberwar', *InformationWeek*, 15 January 2004, wherein he cites a report from David Fraley of Gartner which noted that nations would be able to carry out cyber-warfare by 2005, available at <http://www.informationweek.com/story/showArticle.jhtml? articleID=17301666>, accessed 3 March 2008.

[52] Antone Gonsalves, 'Gartner: Dependence On Internet Boosts Risks of Cyberwar', *InformationWeek*.

[53] Antone Gonsalves, 'Gartner: Dependence On Internet Boosts Risks of Cyberwar', *InformationWeek*.

[54] Drew Clark, 'Computer security officials discount chances of "digital Pearl Harbor"', *National Journal's Technology Daily*, 3 June 2003, available at <http://www.govexec.com/dailyfed/0603/060303td2.htm>, accessed 3 March 2008.

[55] Clark, 'Computer security officials discount chances of "digital Pearl Harbor"'.

[56] 'Behind the Firewall—The Insider Threat', 15 April 2003, ARTICLE ID: 2122. See <http://enterprisesecurity.symantec.com/article.cfm?articleid=2122&PID=14615847&EID=389>.

[57] 'Behind the Firewall—The Insider Threat'.

[58] 'Behind the Firewall—The Insider Threat'.

[59] Clark, 'Computer security officials discount chances of "digital Pearl Harbor"'. I must also thank Richard Hunter for including me in the Gartner Research work of 2003.

[60] David McGlinchey, 'Agencies, Congress urged to upgrade computer security planning', GovExec.com, Washington DC, 17 March 2004, available at <http://www.govexec.com/dailyfed/0304/031704d1.htm>, accessed 3 March 2008.

[61] 'Fighting the worms of mass destruction'.

[62] 'Fighting the worms of mass destruction'.

[63] 'Fighting the worms of mass destruction'.

[64] International Institute for Strategic Studies, *International Institute for Strategic Studies (IISS) Strategic Survey 2003/4*, Oxford University Press, Oxford, May 2004, p. 51.

[65] 'Fighting the worms of mass destruction'.

[66] 'Fighting the worms of mass destruction'. Gerhard Eschelbeck of Qualys is cited in the article.

[67] See Chris Jenkins, 'Internet Terrorism Fears as Virus Hits', *Australian*, 28 January 2004, p. 3.

[68] Jenkins, 'Internet Terrorism Fears as Virus Hits', *Australian*, 28 January 2004, p. 3.

[69] Jenkins, 'Internet Terrorism Fears as Virus Hits', *Australian*, 28 January 2004, p. 3.

[70] International Institute for Strategic Studies, *IISS Strategic Survey 2003/4*, p. 62.

[71] 'Cert/CC Statistics 1998-2005', Carnegie Mellon Software Engineering Institute, undated.

[72] 'Fighting the worms of mass destruction', citing Brightmail, the world's market leader in filtering e-mails.

[73] 'Fighting the worms of mass destruction', citing Brightmail.

[74] 'Fighting the worms of mass destruction', citing Brightmail.

[75] 'Fighting the worms of mass destruction'. Alan Nugent, the chief technologist at the software company Novell, is cited in the article.

[76] 'Fighting the worms of mass destruction', citing Alan Nugent.

[77] Edward Waltz, *Information Warfare: Principles and Operations*, Artech House Publications, Boston and London, 1998, p. 157.

[78] Waltz, *Information Warfare: Principles and Operations*, p. 157.

[79] Waltz, *Information Warfare: Principles and Operations*, p. 157.

[80] Waltz, *Information Warfare: Principles and Operations*, p. 160.

[81] Frank J. Cilluffo and J. Paul Nicholas, 'Cyberstrategy 2.0', *Journal of International Security Affairs*, No. 10, Spring 2006, available at <http://www.securityaffairs.org/issues/2006/10/cilluffo_nicholas.php>, accessed 3 March 2008.

[82] Department of the Prime Minister and Cabinet, *Protecting Australia Against Terrorism 2006: Australia's National Counter-Terrorism Policy and Arrangements*, Department of the Prime Minister and Cabinet, Canberra, 2006, p. 60, available at <http://cipp.gmu.edu/archive/Australia_ProtectAU Terrorism_2006.pdf>, accessed 3 March 2008.

[83] Department of the Prime Minister and Cabinet, *Protecting Australia Against Terrorism 2006*, p. 61.

[84] Department of the Prime Minister and Cabinet, *Protecting Australia Against Terrorism 2006*, p. 61.

[85] Department of the Prime Minister and Cabinet, *Protecting Australia Against Terrorism 2006*, p. 62.

[86] Heinrich de Nysschen, 'Homeland Security', p. 37.

[87] Paul Kurtz, the executive director of the Cyber Security Industry Alliance, was quoted in Heather Greenfield, 'Industry Officials Sketch Priorities for DHS Cyber Czar', *National Journal's Technology Daily*, 2 October 2006, available at <http://www.govexec.com/dailyfed/1006/100206tdpm1.htm>, accessed 3 March 2008.

[88] Heather Greenfield, 'Industry Officials Sketch Priorities for DHS Cyber Czar'. The article cites Paul Kurtz as quoting this figure.

[89] Department of the Prime Minister and Cabinet, *Protecting Australia Against Terrorism 2006*, p. 45.

[90] Parliament House News Release by the Attorney-General, The Hon Philip Ruddock MP, *Protecting Australia's Critical Infrastructure*, 11 May 2004, available at <http://ag.gov.au/www/agd/rwpattach.nsf/VAP/(CFD7369FCAE9B8F32F341DBE097801FF)~MR03CriticalInfrastructure07 May04.doc/$file/MR03CriticalInfrastructure07May04.doc>, accessed 3 March 2008.

[91] Department of the Prime Minister and Cabinet, *Protecting Australia Against Terrorism 2006*, pp. 45–46.

[92] Department of the Prime Minister and Cabinet, *Protecting Australia Against Terrorism 2006*, p. 46.

[93] Department of the Prime Minister and Cabinet, *Protecting Australia Against Terrorism 2006*, p. 45.

[94] See Mark Davis, 'Canberra, CEOs extend forum on Terrorism', *Australian Financial Review*, 24 June 2004, p. 3.

[95] Davis, 'Canberra, CEOs extend forum on Terrorism', *Australian Financial Review*, p. 3.

[96] The origin of TISN lies with the announcement by the Prime Minister in November 2001 of his desire to form a Business–Government Task Force on Critical Infrastructure. The Task Force, which met in March 2002, recommended the establishment of an information-sharing network which led to the Trusted Information Sharing Network (TISN) for Critical Infrastructure Protection (CIP). In December 2002, the Council of Australian Governments (COAG) endorsed the development by the National Counter-Terrorism Committee (NCTC) of guidelines for critical infrastructure protection, including the establishment of criteria to identify critical infrastructure and the outlining of appropriate security measures. The TISN was launched at the National Summit on Critical Infrastructure Protection on 2 April 2003. For more details on TISN, see The Department of the Prime Minister and Cabinet, *Protecting Australia Against Terrorism 2006*, p. 47.

[97] Department of the Prime Minister and Cabinet, *Protecting Australia Against Terrorism 2006*, p. 61.

[98] Department of the Prime Minister and Cabinet, *Protecting Australia Against Terrorism 2006*, p. 46.

[99] Ruddock, *Protecting Australia's Critical Infrastructure*.

[100] CSIAAG brings together owners and operators of Australia's communications sector infrastructure in a trusted forum.

[101] This group advises CIAC on cyber-aspects of critical infrastructure protection.

[102] See Australian Homeland Security Research Centre, *National Security Briefing Notes—Advancing domestic and national security practice: 2007 E-Security National Agenda*, July 2007, available at http://www.homelandsecurity.org.au/files/ 2007_e-security_agenda.pdf>, accessed 3 March 2008.

[103] See The White House, *The National Strategy to Secure Cyberspace*, US White House, Office of the Press Secretary, February 2003, available at <http://www.whitehouse.gov/pcipb/cyberspace_strategy.pdf>, accessed 3 March 2008.

[104] These are listed as tasks for the Secretary of the US DHS in *Critical Infrastructure Identification, Prioritization, and Protection*, Homeland Security Presidential Directive No. 7, US White House, 17 December 2003, available at <http://www.whitehouse.gov/news/releases/2003/12/20031217-5.html>, accessed 3 March 2008.

[105] Drawn from discussion in Cilluffo and Nicholas, 'Cyberstrategy 2.0', and the *US National Strategy to Secure Cyberspace*.

[106] Drawn from *Critical Infrastructure Identification, Prioritization, and Protection*, Homeland Security Presidential Directive No. 7.

[107] I acknowledge the contribution of my colleague Brigadier Steve Ayling for his thoughts on the framework.

[108] See also Accenture, *The Accenture Security Practice: Security and the High-Performance Business*, 2003, available at <http://whitepapers.silicon.com/0,39024759,60086441p,00.htm>, accessed 3 March 2008.

[109] I am indebted to my colleagues Philip Dean and Bruce Talbot for their assistance in clarifying my thinking of how a trusted information infrastructure could be developed.

[110] For an insight into more effective joint fires for the future, see Alan Titheridge, Gary Waters, and Ross Babbage, *Firepower to Win: Australian Defence Force Joint Fires in 2020*, Kokoda Paper no. 5, The Kokoda Foundation, Canberra, October 2007.

[111] Donald Reed refers to this as the 'new strategic reality' in Donald J. Reed, 'Why Strategy Matters in the War on Terror', *Homeland Security Affairs*, vol. II, no. 3, October 2006, p. 5, available at <http://www.hsaj.org/pages/volume2/issue3/ pdfs/2.3.10.pdf>, accessed 3 March 2008.

[112] Reed, 'Why Strategy Matters in the War on Terror', *Homeland Security Affairs*, p. 13. Reed sees this as the essence of strategy going forward.

[113] National Commission on Terrorist Attacks upon the United States, *The 9/11 Commission Report*, W.W. Norton & Company, Inc., New York, 2004, p. 363, available at <http://www.9-11commission.gov/report/911Report.pdf>, accessed 3 March 2008.

[114] David Potts (ed.), *The Big Issue: Command and Combat in the Information Age*, Strategic and Combat Studies Institute Occasional Paper no. 45, CCRP Publication Series, February 2003, pp. 244–45, available at <http://www.dodccrp.org/files/Potts_Big_Issue.pdf>, accessed 3 March 2008.

Chapter 6

An Australian Cyber-warfare Centre

Desmond Ball

Introduction

The Australian Defence Force (ADF) is in the process of being transformed to enable it to gain information superiority in future contingencies.[1] Substantial elements of its future Information Warfare (IW) architecture are already in place, such as the *Collins*-class submarine, some of the satellite communications (SATCOM) systems, and some of the land-based intelligence facilities, but these will all have to be extensively modified and technically updated. However, most of the ADF's force and support elements remain inadequately networked. Other advanced capabilities, including the Royal Australian Navy (RAN)'s Air Warfare Destroyers (AWDs), new Royal Australian Air Force (RAAF) fighter aircraft, and various sorts of Unmanned Aerial Vehicles (UAVs) are in the process of acquisition. Complete functional integration of the 'system of systems' remains embryonic. Some essential elements of the ADF's future IW architecture are barely conceived. The most important is a cyber-warfare centre and its operational capabilities.

There has been considerable development of doctrine by the ADF since the early 2000s. In June 2002, the ADF released its doctrinal statement on Australia's approach to warfare. This occurred at about the same time that the notions of being able to gain an information advantage, dispersing forces, and networking them began to appear. The ADF argued that its aim for the future was to obtain common and enhanced battlespace awareness and, with the application of that awareness, deliver maximum combat effect. It would seek to achieve this through networked operations, which would necessitate a comprehensive 'information network' that would link sensors (for detection), command and control (C2) (for flexible, optimised decision-making), and engagement systems (for precision application of force).[2]

In 2002, the ADF also released *Force 2020*—the ADF's vision statement—whereby networked operations were seen as allowing the war-fighter, through superior command decision-making supported by information technologies coupled with organisational and doctrinal agility, to utilise relatively small forces to maximum effect. *Force 2020* states that 'in the force of 2020, we will have transitioned from platform-centric operations to Network-Enabled

Operations'. The aim is 'to obtain common and enhanced battlespace awareness, and with the application of that awareness, deliver maximum combat effect'. Shared information 'allows a greater level of situational awareness, coordination, and offensive potential than is currently the case'.[3] Further doctrinal development proceeded through 2003, including production of an NCW Concept Paper and an *NCW Roadmap*, which were completed by December 2003, and promulgated in February 2004.[4]

However, the work in 2002–2003 was fundamentally incomplete. It was mostly concerned with enhancing and sharing battlefield awareness and with shortening decision cycles; it essentially ignored the offensive opportunities and challenges of Network Centric Warfare (NCW), and the offensive role of IW more generally.

The *NCW Roadmap*, released in February 2007, reflected enormous recent progress. It articulated a plan for managing the transition of the ADF 'from a network-aware force to a seamless, network-enabled, information-age force', and provided a series of 'milestones' that 'the ADF views as critical to the realisation of its vision for NCW'. It described the mechanisms through which NCW considerations, such as cost, connectivity and vulnerability issues, are now addressed in the capability development process. And it stresses that the purpose of NCW is to enhance the ADF's 'warfighting effectiveness', specifically mentioning 'the offensive support system'.[5]

Since September 2001, a major focus of network-centric activity, across the whole–of-government, has concerned counter-terrorism. The 'war on terror' has stimulated some aspects of Information Operations (IO) while distracting planners from the longer-term construction of an all-embracing NCW architecture. The Defence Signals Directorate (DSD) has enhanced its capabilities for monitoring and tracking mobile phones, and for surveilling websites, Internet usages and international email traffic. It played a key role in the capture of the organisers of the Bali bombings in October 2002. Imam Samudra was arrested in November 2002 after sending an email. Mukhlas was traced by his mobile phone, even though he changed his Subscriber Identity Module (SIM) card every two days and spoke for only a few seconds at a time. Azahari bin Husin, who also helped plan the second Bali bombings in October 2005, was killed in a shoot-out with police in November 2005 after DSD monitored and tracked the mobile phone of one of his accomplices.[6] However, these achievements have been essentially defensive, involving investigative and forensic activities, rather than exploiting cyber-space for offensive IO.

Fundamental issues concerning the development of NCW capabilities remain unresolved, while the role and place of offensive IO and an institutionalised cyber-warfare centre are yet to even be considered. There is a palpable risk that Australia will be caught deficient in some critical capability necessary for securing our most vital national interests in the security environment of the late 2010s and the 2020s.

There is a myriad of complex and extremely difficult issues that require resolution before radically new C2 arrangements can be organised, new technical capabilities acquired and dramatically different operational concepts tested and codified. These include the extent to which complete digitisation and networking of the ADF will permit flatter C2 structures; the availability of different sorts of UAVs and the timeframes for their potential acquisition; the role of offensive operations and the development of doctrine and operational concepts for these; the promulgation of new Rules of Engagement (ROE); and a plethora of human resource issues, including the scope for the creative design and utilisation of reserve forces and other elements of the civil community.[7] These matters will take many years to resolve and even longer, in some cases at least a decade, for the ensuing decisions to be fully implemented.

Numerous organisations currently have important responsibilities concerning some aspect of network vulnerabilities, security and regulation, in addition to the Department of Defence, the ADF and the intelligence and security agencies. Defence has a responsibility to defend vital national infrastructure, and is critically dependent on parts of that infrastructure for command, operational and logistical activities, but it is not the lead agency with respect to network security. Within Defence and the ADF, NCW-related activities remain compartmented; between Defence and the other national authorities, the coordination is fitful, sectoral and poorly organised for IO. No agency appears to be responsible for planning and conducting offensive cyber-warfare.

This chapter proposes the establishment of an Australian Cyber-warfare Centre. It argues that a Centre of some form is necessary to provide coordination of all matters concerning cyber-warfare in Australia. It would provide an institutionalised agency for the development of doctrine, operational concepts and ROE for the ADF concerning cyber-warfare. It would provide a mechanism for ensuring not only that all proposed new Defence capabilities are optimised with respect to comprehensive networking, but also that the requirements for future cyber-activities are satisfactorily identified and articulated. It would develop cyber-warfare contingency plans and specify preparatory actions. It would plan and conduct offensive as well as defensive operations. This chapter also considers several associated issues, including issues involved in determining the best location for such a Centre; the implications of cyber-warfare for command arrangements, intelligence processes and covert activities; and the need to develop appropriate ROE, doctrine and operational concepts. It also includes a brief summary of some regional developments with respect to the institutionalisation of cyber-warfare activities and the practice of cyber-warfare techniques. Finally, it argues that the establishment of an Australian

Cyber-warfare Centre has now become a matter of considerable urgency, and that without it Australia will lack a core system in its 'system of systems' required for warfare in the thoroughly networked Information Age.

The relevant organisations and their coordination

The overall responsibility for e-security in Australia rests with the Attorney-General's Department, through its charter to protect the National Information Infrastructure (NII), which 'comprises information systems that support the telecommunications, transport, distribution, energy, utilities, banking and finance industries as well as critical government services including defence and emergency services'.[8] Its mission is essentially defensive, primarily identifying and coordinating responses to incidents that seriously affect the NII.

The government's 'core policy development and coordination body on e-security matters' is the E-Security Coordination Group (ESCG), established in 2001. It is the 'lead agency addressing e-security matters'.[9] The ESCG is supported by the Critical Infrastructure Protection Group (CIPG), which is responsible for 'identifying and providing advice on the protection of Australia's information infrastructure where the consequences of a security incident are defined as critical'. It 'evaluates the threats and vulnerabilities to the NII', and coordinates crisis management arrangements with other Commonwealth agencies, including with respect to 'defence, national security and counter-terrorism programs'. It is chaired by the Attorney-General's Department, and includes representatives from the Australian Federal Police (AFP), which provides 'an enhanced law enforcement response capability'; the Australian Security Intelligence Organisation (ASIO), which provides intelligence analysis and threat assessment advice; the DSD, which provides 'enhanced incident analysis and response for Commonwealth agencies'; and the Australian Securities and Investments Commission, which undertakes 'detection, investigation and prosecution of electronic fraud in the financial sector'.[10]

There are many agencies within the Department of Defence and the ADF concerned with some aspect of NCW, including monitoring of the electro-magnetic spectrum and cyber-space; ensuring information security (Infosec) and e-security with respect to both national and Defence information and communications systems; conducting research, development and testing of NCW concepts and equipment; and addressing NCW criteria in the capability development process.

The Director General Capability Plans (DGCP) 'provides integration and coordination of NCW with other capability development matters'. The Director General Integrated Capability Development (DGICD) 'provides cross-project NCW integration'. The Director of NCW Implementation 'provides research and policy support' in NCW matters for the capability planning process. The Network

Centric Warfare Project Office (NCWPO) is 'the battlespace architect'; it is responsible for 'ensuring cross-project integration ... through testing NCW compliance with battlespace architectures'. The Chief Information Officer Group (CIOG) 'manages the Network Dimension of Defence NCW capability'. The Intelligence and Security Group (I&SG) is responsible for development of the intelligence component of Defence NCW capability, and for 'managing the implementation and ongoing development of the *Intelligence, Surveillance and Reconnaissance Roadmap*'.[11]

The DSD, Australia's largest intelligence agency, responsible for both the collection of foreign signals and the security of the national information and communications systems, has extensive capabilities relating to cyber-warfare. It has broadened, with respect to its collection activities, from focusing almost entirely on the interception of information 'in motion', as electro-magnetic waves travel through the ether, to now also undertaking the collection and manipulation of information 'at rest', stored on computer databases, disks and hard drives.[12]

DSD has two stations concerned with intercepting SATCOM in the region, monitoring long-distance telephone calls, emails, facsimiles, and computer-to-computer data exchanges. DSD's largest station, at Shoal Bay, near Darwin, is primarily concerned with intercepting Indonesian communications, including both radio transmissions and SATCOM. Project *Larkswood*, which began in 1979, involves the interception of Indonesian SATCOM, and especially those involving Indonesia's *Palapa* communications satellite system. It also includes the communications of other Association of South East Asian Nations (ASEAN) countries that use the *Palapa* system.[13] Many more dish antennas were installed in the late 1990s, making eleven as at September 1999. Most of the new antennas were designed to intercept various sorts of SATCOM involving Indonesia, including mobile satellite telephone (Satphone) conversations using Inmarsat and Global System for Mobile Communications (GSMC) services.[14] DSD's other SATCOM signals intelligence (SIGINT) station is at Kojarena, near Geraldton, WA; it became operational in the mid-1990s, and currently has five large radomes. It is able to monitor selectively the communications from more than a hundred geostationary satellites stationed along the equator from about 40°E to about 170°W longitude.[15] The station reportedly functions as part of the much-publicised 'Echelon' system.[16]

DSD is also Australia's 'national authority' for Infosec. DSD's Information Security Group is responsible for 'the protection of Australian official communications and information systems', with respect to 'information that is processed, stored or communicated by electronic or similar means'. The Group also works with private industry in relation to the development of new cryptographic products, and evaluates Infosec products for industry.[17]

The ADF maintains a variety of electronic warfare (EW) capabilities which are relevant to cyber-warfare. The RAAF's Electronic Warfare Operational Support Unit (EWOSU) was established in Salisbury, SA, in 1991. One of its first responsibilities was to compile 'the first integrated electronic warfare intelligence data base in Australia'.[18] In 1976, the Australian Army raised 72 Electronic Warfare Squadron at Cabarlah, Qld, the home base of the Army's 7 Signal Regiment, to provide EW support to Army forces. It is equipped with a variety of communications intelligence (COMINT) and EW systems, employed for high-frequency (HF) and very high-frequency (VHF) interception, DF, and jamming operations.[19] During the International Force East Timor (INTERFET) operation in late 1999–2000, a component of the Squadron provided the headquarters in Dili with 'timely indicators and warning', and, 'as a secondary task', provided other reconnaissance, surveillance and intelligence collection services.[20]

EW and cyber-warfare are becoming conflated as the electro-magnetic environment merges with cyber-space. Cyber-techniques will be increasingly used to penetrate the electronic components in weapons systems, collecting electronic intelligence to inform the development of electronic support measures (ESM), electronic countermeasures (ECM) and electronic counter-countermeasures (ECCM). ECM and ECCM operations will involve a conjunction of radio-EW and cyber-attacks.

The Defence Science and Technology Organisation (DSTO) has a major role in the implementation of NCW in the ADF, providing 'essential scientific and technological support' with respect to intelligence, surveillance and reconnaissance (ISR), communications, human-computer interfaces, and decision-support tools. Its work on NCW is coordinated by a NCW Steering Group that was formed in early 2003, and includes the development of technologies for battlespace communications and protection of the infrastructure, as well as integration of future weapons systems into the C2 and engagement grids.[21]

The DSTO has recently initiated a series of 'Net Warrior exercises' to 'build, demonstrate and enhance' ADF battlespace interoperability. The participants include the Airborne Early Warning and Control Aircraft (AEW&C) Testbed, the Air Defence Ground Environment Simulator (ADGESIM), and a 'Future Ship' maritime platform. An important focus has been the tactical data-links for exchanging 'battle-space situational awareness information', including the use of 'Internet-based transmission approaches'.[22]

The ADF Warfare Centre at Williamtown is involved in the development of doctrine and the delivery of specialist courses on joint EW and joint IW.

The AFP has considerable expertise in several important aspects of cyber-warfare. It is capable of processing large quantities of digital imagery, such as recorded by closed-circuit television systems. Its Telecommunications Interception Division, which is particularly skilled in monitoring mobile phones, has expanded substantially since the late 1990s, initially under the National Illicit Drug Strategy (NIDS),[23] and since 2001 in accordance with the Australian Government's counter-terrorism agenda. In the case of the so-called 'Bali nine', the Australian heroin smugglers arrested in Bali in April 2005, AFP personnel reportedly cracked the Personal Identification Number (PIN) codes on some of the 10 mobile phones seized, enabling them to identify the network providers and obtain records of 'every phone call made or received during the life of the cards in each mobile'.[24] It is also able to intercept emails, Short Message Service (SMS) and voicemail messages 'that are temporarily delayed and stored during passage over a telecommunications system'.[25] The Australian High Tech Crime Centre (AHTCC), which is hosted in Canberra by the AFP, provides a 'national coordinated approach to combating serious, complex and multi-jurisdictional' computer-generated crime.[26]

ASIO has a Technical Operations Branch which supports its counter-intelligence and counter-terrorism responsibilities. It has expertise not only in monitoring telephones, but also in covert installation of 'bugs' and other technical devices in embassies, private residences and meeting places, and in penetrating computer-related systems.

The Australian Secret Intelligence Service (ASIS) has a Technical Section which generally conducts technical operations in foreign capitals, although it sometimes cooperates with ASIO in operations against foreign missions in Canberra. For example, it was alleged in May 1995 that ASIS had worked with ASIO to install fibre-optic devices in the Chinese Embassy in Canberra while it was being built in the 1980s.[27] Since the 1960s, ASIS has assisted DSD by obtaining foreign code-books; since the late 1990s, it has also provided DSD with internal telephone and email directories. There has been increasing cooperation between ASIS and DSD since the 1990s with respect to technical collection and surveillance operations in foreign capitals. Offensive cyber-warfare operations, and, indeed IO more generally, will place increasing demands on ASIS for covert support overseas.

The corporate sector, and especially the telecommunications, IT and aerospace companies, is an enormous reservoir of cyber-warfare capabilities. Most of the NII is in private hands. Telecommunications are virtually monopolised by Telstra and Optus. There is a plethora of Internet Service Providers (ISPs), some of them committed to the provision of maximum security for their services, regardless of the implications for access by the authorities. The corporate sector contains technical expertise, entrepreneurial ability and research and development (R&D) capabilities.

Telstra and Optus maintain central parts of Australia's NII. Optus has a new headquarters, with 6 500 staff, at Macquarie Park in northwest Sydney. A Network Operations Centre (NOC) at the headquarters was opened by former Prime Minister John Howard in October 2007. The Optus C-1 communications satellite is particularly critical to the ADF's NCW architecture. Positioned in geostationary orbit over the equator at 156°E longitude (i.e. just north of Bougainville), it provides relatively high data rate links between headquarters and tactical platforms to support current and future C2, surveillance, intelligence, logistics and administrative networks. It carries four Defence payloads (Global Broadcast, ultra high-frequency (UHF), X-band and Ka-band), was successfully launched on 11 June 2003; it allows AEW&C *Wedgetail* aircraft, *Jindalee* Operational Radar Network (JORN) and the ground radar net to share data at required data rates.[28] Optus maintains a Satellite Earth Station at Belrose, which has four 13-metre antennas, one of which is dedicated to controlling the C-1 satellite.[29]

Telstra is the largest provider of local and long-distance telephone services, mobile phone services, and wireless, ADSL and cable Internet access in Australia. It was able to assist DSD during the hunt for the October 2002 Bali bombers. Two Telstra technicians visited Jakarta in late October and spent 'several days at the main link to Indonesia's government-owned telecommunications carrier, Telkomsei', where they extracted 'a database of millions of phone numbers', which was then handed to DSD for processing.[30]

Nearly all the servers and routers used in the Australian NII are made by Cisco Systems, headquartered in California. For example, Cisco provided the Internet Protocol (IP) phones, the wireless local area network (WLAN) and the network security at the new Optus head office in Macquarie Park.[31] Cisco has a Product Security Incident Response Team (PSIRT).[32]

AusCERT, the Australian Computer Emergency Response Team, based in Sydney, is a national agency providing expertise on computer network security, particularly with respect to incident response. It is affiliated with the CERT Coordination Centre in the United States, which studies Internet security vulnerabilities, researches long-term changes in networked systems, and provides information to improve the security of networked systems. AusCERT provides a central point in Australia for reporting on security incidents and dissemination of information relating to threats, vulnerabilities and defensive mechanisms.[33]

The aerospace companies possess a range of R&D, design and manufacturing capabilities directly relevant to the cyber-warfare exercise. These include tactical data-links, C2 systems, antenna and radio frequency (RF) propagation systems, and UAVs, as well as specialist electronic components and testing equipment. There is already extensive cooperation between Defence and many companies with respect to NCW systems. For example, DSTO and ADI Ltd signed an agreement at DSTO's Defence Science Communications Laboratory at Edinburgh, north of Adelaide, in September 2004 to form a 'Strategic R&D Alliance' for the collaborative development of NCW technologies.[34] Raytheon Australia has a test-bed Combat Control System (CCS) at its headquarters in North Ryde in Sydney which can simulate, and test new concepts and connectivities with, the Combat Information Systems (CIS) of both the *Collins*-class submarine and the prospective AWDs.

Research, planning and preparation

The dimensions of the terms of reference for an Australian Cyber-warfare Centre require considerable debate and contemplation, and, indeed, they will eventually only evolve once a Centre has begun operating. However, there are several basic planning functions that would be central in any construct. Its activities would be both defensive and offensive. Indeed, the relationship between these is symbiotic, each nourishing the other. Research into ways of penetrating foreign cyber-systems inevitably uncovers vulnerabilities in Australian systems, while research into possible vulnerabilities often suggests ways of exploiting these for offensive purposes.

A core research function of any Australian Cyber-warfare Centre would be the study of telecommunications architectures—the terrestrial microwave relay networks, SATCOM, and fibre-optic cables—both across the region and in particular countries. SATCOM and microwave relays are reasonably accessible, allowing IPs and *pro formas* for computer-to-computer data exchanges to be identified, and providing opportunities for hacking into command chains, combat information systems, air defence systems and databases. This research activity would also involve the identification of the mobile phone numbers and email addresses of foreign political and military leaders.

Another core research function would be the study of the electronic sub-systems in major weapons systems, such as the avionics of particular combat and support aircraft. This would include, for example, finding ways of penetrating the 'firewalls' protecting avionics systems and of using wireless application protocols (WAPs) to insert 'Trojan horses'. This would conceivably allow Australian cyber-specialists to effectively hijack adversary aircraft (and to choose between hard or soft landings for them). In other cases, it would allow electronic components to be disabled or deceived—essentially conducting ECM and ECCM operations through cyber-space.

A Centre would be centrally concerned with studying the vulnerabilities in both Australian and foreign networks and developments in viruses, worms,

'Trojan horses' and other threats to computer-based systems. Publicly acknowledged vulnerabilities in servers indicate promising routes for exploitation. In June 2001, for example, CERT reported a critical flaw in the Hypertext Transfer Protocol (HTTP) component of Cisco Internetwork Operating System (IOS) software using local authentication databases, which 'allows an intruder to execute privileged commands on Cisco routers' and to effectively take 'complete control' of affected systems.[35] In June 2006, multiple vulnerabilities were reported in certain versions of the Cisco Secure Access Control Server (ACS) for Windows, a key part of Cisco's 'trust and identity management framework' and a cornerstone of its Network Admission Control (NAC) system. Some of the vulnerabilities caused the ACS services to crash, while others allowed 'arbitrary code execution if successfully exploited'.[36]

The study of viruses and worms would be not merely for remedial or longer-term protective purposes, but even more importantly would inform the R&D of superior viruses and 'Trojan horses'—making them more malicious, or more selective, or more difficult to trace and diagnose, or less able to be fixed. Some recent examples are the VBS/Loveletter worm (appearing in 2000 and causing between US$5 and US$10 billion dollars in damage), which used a back-door 'Trojan horse'; the Code Red and Code Red II worms in 2001, which attacked the Index Server in Microsoft Internet Information Servers; the SQL Slammer worm, which attacked vulnerabilities in the Microsoft SQL Server; the Blaster worm, which exploited a vulnerability in Microsoft Windows systems; Sobig and MyDoom worms, which spread rapidly via emails; Witty, which exploited vulnerabilities in several Internet Security Systems (ISS); and Santy, a 'Web-worm' that exploited vulnerabilities in Google.[37] Systematic exploration of all known viruses would suggest the most lucrative avenues to explore.

Destructiveness is not necessarily the objective. Although there is a place in IO for relatively crude cyber-operations, such as defacement of websites and Denial of Service (DS) attacks, the most effective and successful cyber-warfare activities are those in which control of computer-related systems is taken without detection by the hosts. Covert corruption of databases, deception of sensor systems, and manipulation of situational awareness is much more likely to produce favourable strategic and tactical outcomes.

A Cyber-warfare Centre would be responsible for the preparation of contingency plans. These would include the development of various forms of 'Trojan horses' designed to surreptitiously corrupt data and files, and matched to particular national stock exchanges, power utilities, air traffic control systems and other information infrastructure; of plans for disabling and deceiving critical elements of military chains of command; and plans for targeting the computer, communications and electronic systems used by particular individuals and agencies. Scenarios would be continually researched and techniques practised to ensure the currency of the plans in contingent circumstances.

A Cyber-warfare Centre would be responsible for identifying the preparations necessary for expeditious implementation of the plans, including the preparations for offensive operations. Some of this preparatory activity will involve the placement of taps on communications systems, of intercept equipment in microwave alleys, and of various electronic devices on antenna systems and communication junctures in foreign countries—to monitor communications, identify IPs and *pro formas*, collect local electronic emanations for the application of countermeasures, and to manipulate and deceive air defence and logistical systems. Devices could be implanted in radars and other sensor systems, or at junctures in their data-links. It is obviously easier to do this before crises or wars eventuate. A Cyber-warfare Centre would have to work very closely with designated ASIS or Special Forces elements with respect to these sorts of activities.

The proportion of both international and local telecommunications traffic being conveyed by fibre-optic cables has increased rapidly since the late 1980s, notwithstanding the increasing volume of mobile telephony connected by both satellite and terrestrial transponders. A rising proportion of voice telephony is being carried by the Internet, via cable, satellite and wireless, as Voice Over Internet Protocol (VOIP) communications. Current trans-oceanic fibre-optic cables typically have four or eight pair of fibre strands, each pair providing four channels, with a capacity of 10 Gigabits per second per channel. Systems have been demonstrated which can carry 14 Terabits per second (111 Gigabits per 140 channels) over a single optical fibre.[38] However, tapping fibre-optic cables is much more difficult than intercepting satellite or terrestrial microwave communications. It requires considerable expertise and specialised equipment, and direct access to the cables.

There are two approaches to tapping fibre-optic cables. One is to access the amplifier or repeater points which regenerate the signals, and which are typically every 160 km or so. This is relatively easy in older systems, which use opto-electronic repeater amplifiers. These convert the optical signals into electrical signals, clean and amplify them and then convert them back to optical for re-transmission; the signals can be intercepted by external induction collars during their electronic stage.[39]

More modern optical cable systems use Erbium [Er]-Doped Fibre Amplifiers (EDFA), in which the signal is boosted without having to be converted into electricity. At each EDFA repeater point there is a small internal tap that takes signals from the eastwards fibre and sends them back along the westwards fibre to let the cable operators diagnose cable fault points very accurately. These signals can be monitored by inserting tap couplers into the EDFAs, although care must be taken to avoid a voltage drop.

The second approach involves the 'scrape and bend method', in which a small piece of cladding is removed from one side of the cable, allowing a detectable amount of light to escape but not enough to alert the cable operators. The exposed fibre is placed in a special reader unit that slightly bends it so that some of the light is refracted (due to it hitting the glass close enough to the perpendicular), and a photon sensor or light detecting device then reads the escaping light. Dummy light packets may have to be inserted so that photon loss is not noticed.[40]

Transmission of the intercepted data is a formidable problem. A cable can be carrying hundreds of gigabits of data, or the equivalent of a hundred million telephone calls at a time. It requires prioritising, based on careful consideration of future intelligence requirements, as well as placement of equipment at the tap sites. The techniques involved include distinguishing the Synchronous Optical Network (SONET) frames that carry the multiplexed digital traffic; concentrating on selected IPs and other easily sorted packages; and using filters that filter terabits per second down to reasonable data level.

In 2005, the USS *Jimmy Carter*, one of the new *Seawolf* class of submarines, was extensively modified for a range of covert missions, including tapping undersea optical cables.[41] However, these missions are obviously extremely complex as well as very expensive.

Fortunately, signals are rarely conveyed by optical cable, let alone undersea cable, for the whole of their journey from sender to recipient. Undersea cables have landing points where they connect with satellite ground stations, terrestrial microwave relay stations, or other cable systems, which in turn often connect with mobile telephony or broad-band wireless transponders. The terminals, junctions and switching centres, as well as the Network Access Points (NAPs), now usually called Internet Exchange Points (IXPs), which serve as Internet exchange facilities, are more accessible and likely to be more lucrative than most undersea cables.

Offensive activities

Many of the posited functions of a Cyber-warfare Centre are already being performed, to a greater or less extent and with unsatisfactory coordination, by one or more of the organisations operating in the Defence intelligence or cyber-security areas. However, none of them has any mandate for the planning and preparation of offensive cyber-warfare activities. Offensive capabilities, represented by the strike capacity of the F-111s and important parts of the Army and RAN, are an essential feature of Australian strategic policy. They must be complemented by offensive NCW and cyber-warfare capabilities at the operational level.

The ADF has moved slowly to acknowledge the offensive dimension of NCW. *Force 2020*, the vision statement issued in 2002, stated that 'in the force of 2020, we will have transitioned from platform-centric operations to Network-Enabled Operations', and that the objective is 'to obtain common and enhanced battlespace awareness, and with the application of that awareness, deliver maximum combat effect'. It said that shared information 'allows a greater level of situational awareness, coordination, and offensive potential than is currently the case'.[42] This remained the only mention of the word 'offensive' in the public guidance for another five years, although in 2002–2003 the Knowledge Staff under the Chief Knowledge Officer argued that, among its transformational capabilities, NCW would provide 'an offensive information operations capability'.[43]

The 2007 *NCW Roadmap* also argued that a networked force would 'facilitate enhanced situational awareness, collaboration and offensive potential'. It described 'the offensive support system', which is 'predominantly based in the engagement grid [of the NCW architecture]', but which also 'combines aspects of the sensor and command and control (C2) grids while exploiting the information network to exchange information between all the grids'.[44]

However, this conception of offensive activities is essentially limited to Network-enabled operations which exploit networking to provide enhanced situational awareness, more informed targeting, and greater precision in weapons delivery. Indeed, the Knowledge Staff classed offensive NCW not only as 'effectors', but as 'weapons systems'.[45] It does not extend to the conduct of offensive cyber-warfare activities—hacking, disabling information infrastructures, disrupting chains of command and decision-making processes, corrupting databases and conducting sophisticated IW—which could well have at least as much impact on some conflict outcomes as more efficient and more effective application on conventional force.

Information Warfare and the intelligence process

IW, and cyber-warfare in particular, poses several new challenges for the intelligence community. The centrality of intelligence collection and analysis is enhanced. Timeliness becomes even more critical; indeed, analysis and assessment become conflated with operations. New intelligence skills are required.

The intelligence collection and processing stations, the EW centres and the cyber-warfare facilities will effectively be integrated. The intelligence centres disseminate the processed intelligence, collated from all sources (especially SIGINT and imagery intelligence (IMINT)) in real-time to the high command, subordinate headquarters and staffs, and to field units. The EW centres maintain catalogues of electronic order of battle (EOB) data about radar systems and other electronic emitters in prospective areas of operations. This includes data on the location of the emitters, their signal strengths and frequencies, the pulse width and pulse length of the signals, and the physical descriptions of the emitting antenna system. The cyber-warfare centre is responsible for both offensive and defensive cyber-activities. It penetrates foreign computer networks, implants viruses, worms and 'Trojan horses', conducts DS attacks, defaces websites, sends misleading information, and disrupts or manipulates connected sensor and information systems.

These centres not only provide intelligence and EW and cyber-warfare capabilities to support the conventional functional and designated commanders; they are also *integrally* involved in the planning and conduct of operations. In future wars (including prospective phases in the 'war on terror'), the winners in the long-term will not necessarily be those who enjoy military success on the battlefields but those who win the information war. In many (but not all) contingencies, the IO units could well play more determinate roles than the conventional force elements. They are the essence of 'effects-based' operations.

NCW and IO have fundamental implications for the role and place of the intelligence process, although this was ignored in the Flood inquiry into Australia's intelligence agencies in 2004.[46] In IO activities, the intelligence process is categorically conflated with the conduct of operations. The role of intelligence changes from a staff agency to an instrumental service. The intelligence cycle becomes the definitive sequence in the operations themselves. In exemplary cases, remotely-controlled sensor systems serve as both intelligence sources and shooters.

This conflation is greatly facilitated by UAVs. Its essence was demonstrated in the use of a *Predator*, armed with *Hellfire* missiles, to hit a car in Yemen on 3 November 2002, killing its six occupants, including the al-Qaeda leader responsible for planning the attack on the USS *Cole* in October 2000.[47] The *Predator* was remotely-piloted from Djibouti, with the surveillance imagery relayed in real-time to a field user equipped with a remote video terminal and to the Central Intelligence Agency (CIA)'s headquarters in Virginia.[48]

The Defence intelligence agencies will have to be drastically reformed, and in parts substantially augmented, in order to perform their central role in NCW and IO/IW. They presently lack important technical capacities, and are surely incapable of providing the timely, accurate and insightful intelligence necessary, when operationalised through IW and cyber-warfare activities, to manipulate the policy-making and decisional processes of notional adversaries.

Command issues

Construction of an IW architecture, including a cyber-warfare component, raises numerous important organisational and command issues for the ADF. Networking will provide more direct sensor-to-shooter connections and enable a more flattened C2 structure. IO generally do not require, and indeed are likely to be impeded by, the erstwhile hierarchical C2 structures of traditional armed forces. With all force elements digitised and connected with sufficient bandwidth, the high command and the 'front-line' IO units can work with a shared battlefield awareness; orders can be issued by the high command to the IO units directly and in real-time, without passing through intermediate echelons; and the IO units, with full appreciation of both the tactical and strategic dimensions, can operate with considerable autonomy from intermediary oversight.

Most IW involve several Services and Defence agencies, with (currently) complex, complicated and distributed command arrangements—as in the plausible IO mission scenario in which a *Collins*-class submarine embarks Special Forces and 'special intelligence' personnel to implant electronic devices in adversary communications, control, command and intelligence (C3I) systems to enable cyber-warriors, using UAVs for connectivity, to penetrate and take control of adversary networks, providing other ADF elements with unrestricted access to adversary telecommunications systems and airspace. Civilians would be integrally involved in the conduct of important operations, especially those involving cyber-warfare. DSTO will accrue new responsibilities, with large elements also directly involved in operations, as the 'wizard war' becomes real-time. New concepts and mechanisms are required for the utilisation of reserves and mobilisation of other civil resources (especially IT and super-computer resources).

Plausible operations could involve a cyber-warfare centre in Canberra working in direct support of air strikes by corrupting an adversary's air defence data, or supporting amphibious lodgements by confusing an adversary's sensor systems and response processes, or supporting counter-terrorist operations by temporarily disrupting electrical power in a particular locality, while depending on UAVs, submarines and Special Forces to provide access to the adversary's networks. Civilians would be involved in the actual conduct of operations, first, where particular hacking skills are required, and, second, to provide subject expertise. Experts on a foreign national financial system, or the key personalities and political processes in a foreign government, would literally sit next to the hackers, providing direction and advice as penetration is achieved, data surreptitiously distorted, effects modulated, and decisions effectively manipulated.

The issue of creating a Commander of Information Operations, at the equivalent level of the present functional or environmental commanders (i.e. 2-star), has been raised elsewhere.[49] The case will undoubtedly become increasingly compelling, as the role of IO not only as an enabler of more effective air, maritime and land operations but also as a determinant of conflict outcomes becomes more apparent, and other benefits with respect to IO planning, doctrine development and capability development become better appreciated. Operational command and control of a cyber-warfare centre would be simplified, as would some of the tasking arrangements involving special operations units and platforms such as the submarines and UAVs.

A premium on *ante-bellum* activities

IO place a premium on the conduct of preparatory activities, inherently involving covert operations in peacetime. It is a process quite different to the planning and logistics preparations that are involved in conventional operations; it is more akin to the craft of the cryptanalyst.

Information Supremacy requires intimate familiarity with the intricacies of an adversary's C2 structure, public media, defence communications systems, sensor systems, and cyber-networks. It requires the maintenance of comprehensive, accurate and completely up-to-date EOB data for the design of ESM and ECM EW systems and application tactics. It requires detailed knowledge of the antenna systems and electronic equipment aboard adversary combatants (such as the avionics in aircraft) and installed in hardened command centres, in order to design EW, 'front-door', 'in-band' high-power microwave (HPM), and cyber-penetration techniques. Cyber-warriors must explore offensive cyber-warfare, in which software is developed for penetrating the firewalls in designated sectors (such as air traffic control and air defence networks); worms, viruses and 'Trojan horses' are developed, and plans are prepared and tested, for the corruption or disablement of websites and databases; and 'back-door' programs are installed in designated computer networks enabling data to be copied from files without detection, or the cyber-warfare centre to take control of the infected computers.

Much of this would be done in the Australian Cyber-warfare Centre itself, using airborne connectivity with the adversary's networks. However, the process would be greatly advantaged by the prior emplacement of various sorts of devices on adversary C3I systems—such as telephone exchanges, microwave relay towers, radar equipment, and even SATCOM ground control stations.

There are questions about Australia's willingness and capacity for engaging in *ante-bellum* activities, especially those involving identifiable penetrations of cyber-networks or physical implants of technical devices.

Rules of engagement, doctrine and operational concepts

The ROE for IW, including cyber-warfare, will have to be very different from those of traditional military operations. New doctrine will have to be promulgated. New 'laws of war' will have to be developed and accorded some international standing.

New ROE are required to accommodate the transient nature of some of the real-time intelligence available to the shooter and to remove intermediary command levels. For example, early in Operation *Enduring Freedom*, a *Predator*

UAV armed with *Hellfire* missiles, on a mission for the CIA, spotted 'a top Taliban leader' entering a building. The aircraft could have taken a shot at the building, but the CIA officials had to seek permission from US Central Command in Tampa, Florida, and by the time the military officials ordered a strike the Taliban leader had fled.[50]

New ROE are also required for those ADF elements that might be engaged in pre-emptive operations, covert operations, and offensive cyber-warfare activities. US military officials have said that clarifying the ROE for initiating computer network attacks has been 'a particularly thorny issue' in the Pentagon, 'due to larger political considerations, particularly to ensure that the attacks do not have any important unintended political ramifications or effects beyond the target'.[51] During the planning of Operation *Iraqi Freedom*, IO officers encountered opposition from 'the Pentagon's legal community', which was worried about the unintended effects of IO.[52] US officials have said that the activities of the *Compass Call* IW aircraft were determined, both preceding and during the war, partly by legal arguments (which determined that 'jamming a sovereign country is an act of war'), and that 'it was very difficult for us to get our hands around what we were authorized to do before the start of hostilities'.[53]

New operational concepts and doctrine will have to be developed for new areas of activity, especially those involving offensive cyber-warfare. For example, doctrine is required to define and exploit the 'intersection of information warfare and air defence suppression'. Special mission aircraft, such as the RC-135 *Rivet Joint*, can tap into enemy radar systems to 'see' what they are detecting—and hence instruct fighters to either press home or abort an attack.[54]

Some of this doctrine and associated ROE will have to be developed in the absence of any relevant international law. The killing of the al-Qaeda operatives by the *Predator* UAV in Yemen on 3 November 2002 raised troubling ethical questions. Swedish Foreign Minister Anna Lindh called it 'a summary execution that violates human rights'.[55]

The inherent transnational and non-State attributes of cyber-activities, confounding distinctions between external and internal security operations, pose not only new technical challenges but also contain new risks, in terms of both national vulnerabilities and threats to civil liberties. These have to be addressed in both national legislation and ROE.

Capability planning

The 2007 *NCW Roadmap* described the mechanisms through which the NCW Concept has now thoroughly infused the capability development process. All major capability proposals are now rigorously vetted from an NCW perspective, primarily focused on their 'level of connectivity and integration requirements', their IT vulnerabilities, and the potential contribution of their Combat

Information Systems to provide enhanced ISR capabilities and to enable more combat effect.[56]

A Cyber-warfare Centre would add a new and critically important dimension to this process. Major capability proposals have to have initiators and mentors. There has to be some place concerned with identification of long-term cyber-warfare requirements apart from critiquing the NCW aspects of other proposals. A Cyber-warfare Centre would provide an institutionalised advocate for funding and specialised equipment beyond the scope of the current process.

Some of the specialised requirements equipment will only become apparent after such a Centre has been functioning for a while. Much of it will consist of assorted miniature devices for implantation at various physical places in adversary networks, but there might also be major support platforms. UAVs offer extraordinary promise for both enhanced and precisely-targetable COMINT collection and penetration of networks exposed during microwave transmissions. The acquisition of a squadron of *Global Hawks* for SIGINT collection is a serious possibility within the next decade. There are programs to produce a version of the *Global Hawk* with a 3000 lb SIGINT payload, including COMINT capabilities. It might well be the case that three *Global Hawks* (with one on a continuous 36-hour station), equipped with various sorts of antenna systems, could provide comparable COMINT coverage to that of the first *Rhyolite* geostationary SIGINT satellites in the 1970s. Other configurations, focused on 'microwave alleys', could provide direct support for interactive cyber-warriors.

The costs of NCW and IO will not be trivial. The bandwidth requirements of NCW and IO are staggering. Advanced communications satellite systems will be necessary, using laser transmission and Internet routing to provide high-bandwidth connectivity.[57] The networks and servers used by Defence for Network-enabled operations have to be completely secured—not only against terrorists and other non-State actors, but also against the cyber-warfare activities of notional regional adversaries. Special operations units will have to be formed for covert activities, such as placement of devices in microwave relay facilities, optical cable networks, switching centres and air defence systems in particular countries. The construction of a Cyber-warfare Centre could well cost more than a billion dollars. However, this would be spread over many years, and could begin quite modestly, with more robust networking of various current activities and capabilities, and direction and coordination provided by a small core.

Location of a Cyber-warfare Centre

Location is a factor of organisational, technical and operational considerations. A Cyber-warfare Centre would be a national asset, serving grand strategy as much as tactical encounters. It would have to respond to direction from, report to, and interact with agencies at several levels. Its central role in operations has to be reflected in ADF command arrangements. Robust connectivity with the rest of the NCW infrastructure and the NII is fundamental. The physical proximity of all its components is not necessary, so long as functional coordination and cooperation can be organised. Networking enables elements to be distributed, across agencies and geographically. There will inevitably be offices in more than one place, as well as out-rider units in high-tech centres such as the Salisbury/Edinburgh area in north Adelaide.

A Cyber-warfare Centre would have at least two new elements that would require accommodation. One comprises the executive, coordination, planning and management functions, which extend across the whole of government. This is the purview of the National Security Division of the Department of the Prime Minister and Cabinet (DPM&C), which has a mandate 'to foster greater coordination of, and a stronger whole-of-government policy focus on, national security' and which has greatly strengthened the whole-of-government approach to counter-terrorism. There are many aspects of cyber-warfare that require coordination at a national level. This includes all cyber-activities aimed at influencing adversary political and strategic agencies and processes. Covert operations in peacetime must be endorsed at this level because of the risks and consequences of possible exposure. In time of war, the whole-of-government approach would be a crucial feature of the simultaneous application and progressive interaction of kinetic and cognitive effects.

The second element is the operational facility, the place where the cyber-warriors would work. Technical intelligence collection stations, EW centres and cyber-warfare capabilities will remain dispersed, but a place devoted to cyber-operations would promote interaction of the specialist personnel in these areas, where close cooperation is not only essential to operational success, but is likely to encourage future technical advances at the interstices.

It is important to have a place where defensive and much-enhanced offensive activities are co-located. Those working on defensive matters need to keep the offensive planners apprised of the avenues they are finding most difficult to protect. Those working on offensive plans should obviously keep the defensive side informed as they discover potential vulnerabilities in national systems while exploring avenues to exploit. The symbiosis should enhance the security, reliability, capacity and endurance of national networks while maximising the potency and perniciousness of Australia's cyber-warfare capabilities against hostile systems.

The question of location is complicated by the decision to locate the Headquarters Joint Operations Command (HQJOC) near Bungendore, 29 km east of Canberra (and about 28 minutes to drive), rather than at HMAS *Harman* or somewhere else close to the Defence complex at Russell Hill. An obvious place would be in or near the DSD building—the main centre for the collection, processing and analysis of intercepted telecommunications, the main repository of certain specialised cyber-skills, the manager of some of the most secure networks in the world, and the national agency responsible for the defensive cyber-warfare mission, protecting the Australian Government's communications and information systems. However, a component element will now have to be located at Bungendore to serve the Chief of Joint Operations (CJOPS), requiring some bifurcation of the Cyber-warfare Centre and very difficult decisions about which capabilities and activities to maintain at Russell Hill and which to repose at Bungendore.

A component element might also be located at HMAS *Harman*, 11.4 km southeast of Russell Hill (and 19 minutes to drive). It hosts the Defence Network Operations Centre (DNOC), the hub of the third largest communications network in Australia after Telstra and Optus, which provides network support for military operations. The operational elements of the DNOC include the Naval Communications Station Canberra (NAVCOMMSTA), which provides UHF satellite services in support of the RAN and other ADF users; the Naval Communications Area Master Station Australia (NAVCAMSAUS) which supports RAN fleet communications; and the Defence Information Systems Communications Element (DISCE), which provides a secure and survivable communications network to support strategic and tactical operations of the ADF and selected Government departments. Under Project JP 2008 (Phase 3F), a new ground station is to be constructed at HMAS *Harman*, together with two new terminals at the Defence communications station at Geraldton, to upgrade 'the entire Australian Defence Satellite Communications Capability (ADSCC) Ground Segment'.[58]

The HQJOC will have a dedicated fibre-optic cable extending 27 km to the DNOC at HMAS *Harman*. Redundancy will be provided by 'four links into the local carrier network'. The facility will also have 'a back-up satellite link'.[59]

Regional developments

Over the past decade or so, responding to either the Revolution in Military Affairs (RMA) or to the challenges and opportunities of the Internet, many countries have established cyber-warfare organisations of some sort or another. Some of them are attached to national intelligence agencies or Defence Ministries, while others function as part of military command structures. The United States has a variety, spawned by the National Security Agency (NSA), the CIA, the Federal Bureau of Investigation (FBI) and the Department of Homeland Security (DHS) in Washington. In 2000, the US Space Command was given responsibility for both Computer Network Defence (CND) and Computer Network Attack (CNA) missions. After the Space Command was merged into the US Strategic Command (USSTRATCOM) in June 2002, several new organisations were established for planning and conducting cyber-warfare, including the Joint Functional

Component Command for Network Warfare (JFCC-NW), responsible for 'deliberate planning of network warfare, which includes coordinated planning of offensive network attack'; the Joint Functional Component Command for Space and Global Strike (JFCC-SGS), which also houses the Joint Information Operations Warfare Center (JIOWC), responsible 'for assisting combatant commands with an integrated approach to information operations'; and the Joint Task Force for Global Network Operations (JTF-GNO), which has responsibility for Department of Defense cyber-security.[60] The US Navy established a Naval Network Warfare Command (NNWC) at Norfolk in Virginia in July 2002. The US Air Force established a new Cyberspace Command at Barksdale Air Force Base in Louisiana in June 2007, 'already home to about 25 000 military personnel involved in everything from electronic warfare to network defence'.[61] IW teams deploy with combatant commands. Interoperability with the US cyber-warfare architecture requires appropriate institutional arrangements on the part of US allies.

Asia has emerged as the 'early proving ground' for cyber-warfare'.[62] This is especially the case in Northeast Asia, where cyber-warfare activities have become commonplace. China has the most extensive and most tested cyber-warfare capabilities, although the technical expertise is very uneven. China began to implement an IW plan in 1995, and since 1997 has conducted several exercises in which computer viruses have been used to interrupt military communications and public broadcasting systems. In April 1997, a 100-member elite corps was established by the Central Military Commission to devise 'ways of planting disabling computer viruses into American and other Western C2 defence systems'.[63] In 2000, China established a strategic IW unit (which US observers have called 'Net Force') designed to 'wage combat through computer networks to manipulate enemy information systems spanning spare parts deliveries to fire control and guidance systems'.[64]

Chinese cyber-warfare units have been very active, although it is often very difficult to attribute activities originating in China to official agencies or private 'netizens'. Since 1999, there have been periodic rounds of attacks against official websites in Taiwan, Japan and the United States. These have typically involved fairly basic penetrations, allowing websites to be defaced or servers to be crashed by DS programs. More sophisticated 'Trojan horse' programs were used in 2002 to penetrate and steal information from the Dalai Lama's computer network.[65] 'Trojan horse' programs camouflaged as Word and PowerPoint documents have been inserted in computers in government offices in several countries around the world.[66] Portable, large-capacity hard disks, often used by government agencies, have been found to carry 'Trojan horses' which automatically upload to Beijing websites everything that the computer user saves on the hard disk.[67] Since the late 1990s, the People's Liberation Army (PLA) has conducted more than 100 military exercises involving some aspect of IW, although the practice has generally exposed substantial shortfalls.[68]

It has recently been reported that Chinese 'cyber-espionage' activities have been conducted against 'key Australian Government agencies'. According to media reports in February 2008, 'Chinese computer hackers have launched targeted attacks on classified Australian Government computer networks', and that China is 'believed to be seeking information on subjects such as military secrets and the prices Australian companies will seek for resources such as coal and iron ore'. The Chinese activities have reportedly prompted an official review of IT security'.[69]

In August 1999, following a spate of cross-Strait attacks against computer networks and official websites in Taiwan, the Minister for National Defense (MND) in Taipei announced that the MND had established a Military Information Warfare Strategy Policy Committee and noted that 'we are able to defend ourselves in an information war'.[70] In January 2000, the Director of the MND's Communication Electronics and Information Bureau announced that the Military Information Warfare Strategy Policy Committee had 'the ability to attack the PRC with 1,000 different computer viruses'.[71] In August 2000, Taiwan's *Hankuang* 16 defence exercise included training in cyber-warfare, in which more than 2000 computer viruses were tested. Two teams of cyber-warriors used the viruses in simulated attacks on Taiwan's computer networks.[72] In December 2000, the MND's Military Information Warfare Strategy Policy Committee was expanded and converted into a battalion-size centre under the direct command of the General Staff Headquarters, and with responsibilities for network surveillance, defence, and countermeasures.[73] In its *2002 National Defense Report*, released in July 2002, the MND for the first time included discussion of 'electronic and information warfare units'. It proclaimed Taiwan's commitment to the achievement of 'superiority [over the PRC] in information and electronic warfare', and it ranked EW and IW ahead of air and sea defence in terms of current MND focus. It specifically cited such threatening developments by the PRC as 'Internet viruses, killer satellites, [and] electromagnetic pulses that could fry computer networks vital to Taiwan's defence and economy'.[74]

Japan was surprisingly laggard about developing cyber-warfare capabilities. In April 1999, faced with a growing problem of cyber-crime (involving offences such as computer-based fraud, on-line sales of illegal drugs, and transmission of pornography), the National Police Agency set up a 'special unit of cyber-sleuths ... who specialise in investigating computer-related crimes and cyber-terrorism'.[75] A 'specialised anti-hacker task force' was set up on 21 January 2000, but it was quickly shown to be impotent. Two days later there began an intense spate of attacks on Japanese government websites, probably triggered by denials by right-wing Japanese that Japanese troops had massacred Chinese civilians when they seized Nanjing in 1937.[76]

In May 2000, Japan announced plans to establish a Research Institute and an operational unit for fighting cyber-terrorism. The announcement was prompted by further sporadic hacking attacks. Some of these involved a 'cyber war between netizens of South Korea and Japan' over Japanese claims to the disputed Tok-do islets.[77] It also followed revelations in March 2000 that the *Aum Shinri Kyo* (Supreme Truth) sect (responsible for the sarin gas attack in the Tokyo subway in March 1995) had written computer software used by police agencies, which had enabled cult members to obtain secret data on police patrol cars, as well as other software which allowed them access to data on the repairs and inspections of several nuclear power plants.[78]

In July 2000, the Japan Defense Agency (JDA)'s[79] *Defense of Japan 2000* acknowledged, for the first time, the threat posed by IW. It noted that 'there is a greater possibility that invasion and tampering with computer systems by hackers will affect our life immensely', that 'a new computer security base will be established', that facilities would be developed for operational evaluation of computer security systems and techniques, and that JDA personnel would be dispatched to the United States to develop computer security expertise. It also noted that JDA officials contribute to the 'Action Plan for Building Foundations of Information Systems Protection from Hackers and Other Cyberthreats' by 'studying measures against hackers and cyber-terrorism'.[80] It was reported in October 2000 that the JDA's 'cyber-squad' was developing software capable of launching anti-hacking and anti-virus attacks and of destroying the computers of hackers trying to penetrate Japan's defence networks.[81]

South Korea has evidently also moved to establish a cyber-warfare capability. The number of attacks on South Korean commercial and government websites increased markedly during 2000 (partly reflecting the 'cyber-war' with Japanese 'netizens'). The South Korean MND and the National Intelligence Service (NIS) both reported during 2000 that the South Korean armed forces should 'prepare for cyber-warfare in the future from enemy countries' and that they should consider establishing 'specialist units for cyber-warfare'.[82] A National Cyber Security Center attached to the NIS was functioning by 2004.[83]

Even North Korea, the most backward country in East Asia in IT terms, reportedly set up a cyber-warfare unit in the late 1980s. Media reports actually refer to two different places, but these may be different elements of the one agency. An electronic communications monitoring and computer hacking group from the State Security Agency is reportedly located at the Korea Computer Centre in Pyongyang.[84] The North Korean Army created a dedicated cyber-warfare unit, called Unit 121, in 1998. Its staff is estimated to include from 500 to more than 1000 'hackers'. Its capabilities include 'moderately advanced

Distributed Denial of Service (DDS) capability' and 'moderate virus and malicious code capabilities'. In October 2007, North Korea tested a 'logic bomb' containing malicious code designed to be executed should certain events occur or at some pre-determined time; the test led to a UN Security Council (UNSC) resolution banning sales of mainframe computers and lap-top personal computers (PCs) to North Korea.[85] North Korea also uses cyber-space extensively for its propaganda or psychological warfare campaigns.[86]

In Southeast Asia, Singapore has both the leading IT industries and the most advanced cyber-warfare capabilities. Singapore's defence hierarchy 'is committed to the development of an offensive cyber-warfare capability'.[87] The Ministry of Defence and the Singapore Armed Forces initiated a Cyberspace Security Project in the mid-1990s to develop 'countermeasures which respond automatically to attacks on their computer systems'.[88] A dedicated cyber-warfare unit is thought to have been established within the Ministry of Defence, and methods for inserting computer viruses into other countries' computer networks have been developed.[89]

This is not the place to evaluate these regional agencies. They include many different sorts or organisations with wide-ranging responsibilities, not all of them necessarily relevant to Australia's circumstances. They operate in secret. Little is publicly known about them, and this is suffused with misinformation and disinformation. However, they have each accumulated experiences of one sort or another, developed practical and forensic skills, acquired equipment, and undertaken operations with counterpart civilian or military authorities to a greater or lesser extent. This accrual derives from bureaucratic institutionalisation and provides a basis from which 'asymmetric' surprises can be launched. They can only be systematically monitored and countered in institutionalised fashion.

Conclusion

The lack of a Net-war or Cyber-warfare Centre is becoming a critical deficiency in Australia's evolving architecture for achieving and exploiting Information Superiority and Support (IS&S) beyond around 2020. Australia has a plethora of organisations, within and outside Defence, concerned with some aspect of cyber-warfare (including network security), but they are poorly coordinated and are not committed to the full exploitation of cyber-space for either military operations or IW more generally. A dedicated Cyber-warfare Centre is fundamental to the planning and conduct of both defensive and offensive IO. It would be responsible for exploring the full possibilities of future cyber-warfare, and developing the doctrine and operational concepts for IO. It would study all viruses, DS programs, 'Trojan horses' and 'trap-door' systems, not only for defensive purposes but also to discern offensive applications. It would study the firewalls around computer systems in military high commands and headquarters in the region, in avionics and other weapons systems, and in telecommunications centres, banks and stock exchanges, ready to penetrate a command centre, a flight deck or a ship's bridge, a telephone or data exchange node, or a central bank at a moment's notice, and able to insert confounding orders and to manipulate data without the adversary's knowledge. It would identify new capability requirements, including new systems and support platforms for accessing adversary IPs and computer-to-computer *pro formas*. It would task special operations units both for covert preparatory missions in peacetime and during the conduct of offensive IO in conflict situations.

The 2007 *NCW Roadmap* reflected substantial progress with the institutionalisation of NCW perspectives within Defence; however, it also showed, at least implicitly, that vitally important activities are inadequately attended by current structures and processes, and that some sort of Cyber-warfare Centre is best able to address these potentially debilitating deficiencies. Networked databases are useless if the data can be corrupted, providing confusing or misleading information, or if decision-makers lose confidence in them. Expansive networking, incorporating more databases, involving more carriers, and connecting with many more customers, can increase network vulnerabilities; there are more access points for hostile intruders, and more data-links that can be disrupted. Shared and enhanced situational awareness is a superlative 'force multiplier', but it can be disastrous if it is subject to surreptitious manipulation.

The current capability development process ensures that key NCW criteria are examined with respect to all prospective acquisitions, including the levels of connectivity and security, and their contribution to ISR missions, thus increasing the potency of new acquisitions. However, there is no endorsed vision of any notional ADF IW architecture for the period beyond 2020 which can ensure that the sorts of capabilities currently being acquired or being proposed for acquisition are the optimal components of that architecture; there is no agency committed to ensuring that all the requisites for effective cyber-warfare (including equipment) will be in place (which will only become apparent through the plans and activities of some Cyber-warfare Centre).

Furthermore, the 2007 *NCW Roadmap* portrays a severely delimited concept of offensive cyber-operations. It alluded to the central place of the 'offensive support system' in the ADF's NCW architecture. However, its scope is essentially confined to enablement of increases in the combat power of ADF units (through enhanced situational awareness, speedier decision-making, and more precise and more tailored application of force). There is no evident appreciation of the role of offensive cyber-warfare in influencing conflict outcomes quite apart from (but carefully coordinated with) enhanced combat power.

The establishment of an Australian Cyber-warfare Centre has now become a matter of considerable urgency. It is essential for it to be established soon to ensure that Australia will have the necessary capabilities for conducting technically and strategically sophisticated cyber-warfare activities by about 2020.

Several basic issues require considerable further debate, including the best location for a Centre, its organisation and staffing arrangements, its core functions and initial terms of reference, and its ADF command relationships. Some matters will only be resolved once the Centre has been functioning for several years, including some of the command relationships and some of the specialised equipment requirements. Assuming a decision to establish an Australian Cyber-warfare Centre was to be made by 2010, an initial operational capability could be assembled within a couple of years; however, it would not be able to perform all of its ascribed functions, especially those that require the development or acquisition of new capabilities and/or the placement of assorted devices overseas, much before about 2018. This means that an informed and vigorous debate on these issues should be encouraged as soon as possible.

ENDNOTES

[1] Gary Waters and Desmond Ball, *Transforming the Australian Defence Force (ADF) for Information Superiority*, Canberra Papers on Strategy and Defence no. 159, Strategic and Defence Studies Centre, The Australian National University, Canberra, 2005.

[2] Department of Defence, *The Australian Approach to Warfare*, Department of Defence, Canberra, June 2002, available at <http://www.defence.gov.au/ publications/taatw.pdf>, accessed 4 March 2008. See also Waters and Ball, *Transforming the Australian Defence Force (ADF) for Information Superiority*.

[3] Department of Defence, *Force 2020*, Department of Defence, Canberra, 2002, p. 19, available at <http://www.defence.gov.au/publications/f2020.pdf>, accessed 4 March 2008.

[4] Australian Defence Headquarters, *Enabling Future Warfighting: Network Centric Warfare*, ADDP-D.3.1, Australian Defence Headquarters, Canberra, February 2004, available at <http://www.defence.gov.au/strategy/fwc/documents/ NCW_Concept.pdf>, accessed 4 March 2008.

[5] Director General Capability and Plans, *NCW Roadmap 2007*, Defence Publishing Service, Canberra, February 2007, available at <http://www.defence.gov.au/capability/ncwi/docs/2007NCW_Roadmap.pdf>, accessed 4 March 2008.

[6] Cameron Stewart, 'Telstra Operation Helped Track Down Bali Bombers', *Australian*, 7 October 2006, p. 8, on-line version entitled 'Telstra Secretly helped Hunt Bali Bombers' at <http://www.news.com.au/story/0,23599,20537904-2,00.html>, accessed 4 March 2008; and Martin Chulov, 'A Win Against Terror', *Australian*, 7 October 2007, p. 17, available at <http://www.theaustralian.news. com.au/story/0,20867,20536940-5001561,00.html>, accessed 4 March 2008.

[7] Waters and Ball, *Transforming the Australian Defence force (ADF) for Information Superiority*, pp. 61–68.

[8] 'Budget 2001-2002 Fact Sheet. Protecting the National Information Infrastructure: Part of the Government's E-security Initiative', Attorney-General's Department.

[9] Attorney-General, Minister for Communications, Information Technology and the Arts, and Minister for Defence, 'Security in the Electronic Environment', Joint News Release, 27 September 2001; and 'Budget 2001-2002 Fact Sheet. Protecting the National Information Infrastructure: Part of the Government's E-security Initiative', Attorney-General's Department.

[10] Attorney-General, Minister for Communications, Information Technology and the Arts, and Minister for Defence, 'Security in the Electronic Environment'; and 'Budget 2001-2002 Fact Sheet. Protecting the National Information Infrastructure: Part of the Government's E-security Initiative'.

[11] Director General Capability and Plans, *NCW Roadmap 2007*, p. 19.

[12] James Bamford, *Body of Secrets: How America's NSA and Britain's GCHQ Eavesdrop on the World*, Century, London, 2001, p. 480.

[13] Desmond Ball, *Australia's Secret Space Programs*, Canberra Papers on Strategy and Defence no. 43, Strategic and Defence Studies Centre, The Australian National University, Canberra, 1988, chapter 3.

[14] Desmond Ball, 'Silent Witness: Australian Intelligence and East Timor', in Richard Tanter, Desmond Ball and Gerry van Klinken, *Masters of Terror: Indonesia's Military and Violence in East Timor*, Rowman & Littlefield, New York, 2006, pp. 177–201.

[15] Ball, *Australia's Secret Space Programs*, chapter 4.

[16] Jeffrey Richelson, 'Desperately Seeking Signals', *Bulletin of the Atomic Scientists*, vol. 56, no. 2, March/April 2000, pp. 47–51.

[17] 'Infosec', Defence Signals Directorate, available at <http://www.dsd.gov.au/infosec/>, accessed 4 March 2008.

[18] See Desmond Ball, *Signals Intelligence in the Post-Cold War Era: Developments in the Asia-Pacific Region*, Institute of Southeast Asian Studies, Singapore, 1993, p. 83.

[19] Ball, *Signals Intelligence in the Post-Cold War Era: Developments in the Asia-Pacific Region*, p. 83.

[20] Major John Blaxland, 'On Operations in East Timor', *Australian Army Journal*, 2000, pp. 7, 9.

[21] Defence Science and Technology Organisation, 'Network-Centric Warfare', available at <http://www.dsto.defence.gov.au/research/4051/page/4387/>, accessed 4 March 2008. See also Tim McKenna, Terry Moon, Richard Davis and Leoni Warne, 'Science and Technology for Australian Network-Centric Warfare: Function, Form and Fit', *ADF Journal*, no. 17, pp. 62–75.

[22] Arthur Filippidis, Tan Doan and Brad Tobin, 'Net Warrior—DSTO Battlelab Interoperability', Simulation Industry Association of Australia, June 2007, available at <http://www.siaa.asn.au/simtect/2007/Abstracts/70.html>, accessed 4 March 2008.

[23] 'Increased Telephone Interception Capacity', in Australian Federal Police, *National Illicit Drug Strategy Initiatives, November 1997—April 2001* (Second edition), p. 13, available at <http://www.afp.gov.au/__data/assets/pdf_file/6634/nids.pdf>, accessed 4 March 2008.

[24] Philip Cornford and Rob O'Neill, 'Bali Nine Phone Cards Cracked', *Age*, 4 May 2005.

[25] 'Telecommunications Interception Law Dispute Shows Law Needs Overhaul', *Electronic Frontiers Australia*, 31 March 2004, available at <http://www.efa.org.au/Publish/PR040331.html>, accessed 4 March 2008.

[26] See Australian High Tech Crime Centre website at <http://www.ahtcc.gov.au/about_us/index.htm>, accessed 4 March 2008.

[27] Robert Milliken, 'Canberra Acts to Keep an Eye on its Spies', *Independent* (London), 2 June 1995, available at <http://findarticles.com/p/articles/mi_qn4158/ is_19950602/ai_n13986087>, accessed 4 March 2008.

[28] Trevor W. Mahony, 'A Hybrid Civilian/Military Payload to Support Battlefield Communications', *Journal of Battlefield Technology*, vol. 1, no. 1, March 1998, pp. 29–32.

[29] 'Optus Positions for National Satellite Success', December 2001, available at <http://www.optus.net.au/portal/site/aboutoptus/menuitem.813c6f701cee5a14f0419f108c8ac7a0/?vgnextoid=a7ab8336054f4010VgnVCM1000009fa87c0aRCRD&vgncxtchannel=b93cfaf924954010VgnVCM10000029a67c0aRCRD&vgnextfmt=default>, accessed 4 March 2008.

[30] Chulov, 'A Win Against Terror'.

[31] CISCO, 'Optus Charts Future with Cisco Service Oriented Network at Macquarie Park Campus', 19 October 2006, available at <http://newsroom.cisco.com/dlls/global/asiapac/news/2006/pr_10-19.html>, accessed 4 March 2008.

[32] 'CISCO Security Advisories', available at <http://www.cisco.com/en/US/products/products_security_advisories_listing.html>, accessed 4 March 2008.

[33] See the AusCERT website at <http://www.auscert.org.au/>, accessed 4 March 2008.

[34] 'DSTO and ADI Forge New Links in Network Centric Warfare', Defence Science and Technology Organisation, 2 September 2004, available at <http://www.dsto.defence.gov.au/news/3283/>, accessed 4 March 2008.

[35] CERT, 'CERT Advisory CA-2001-14 Cisco IOS HTTP Server Authentication Vulnerability', 28 June 2001, available at <http://www.cert.org/advisories/CA-2001-14.html>, accessed 4 March 2008.

[36] Kevin McLachlan, 'Flaw Found in Cisco Secure Access Control Server', 26 June 2006, available at <http://www.crn.com/it-channel/189601708>, accessed 4 March 2008; and 'Multiple Vulnerabilities in Cisco Secure Access Control Server', 7 January 2007, available at <http://www.securiteam.com/securitynews/5DP0420KAG.html>, accessed 4 March 2008.

[37] 'Timeline of Notable Computer Viruses and Worms', Wikipedia, available at <http://en.wikipedia.org/wiki/Timeline_of_notable_computer_viruses_and_ worms>, accessed 4 March 2008.

[38] Frank W. Kerfoot and William C. Marra, 'Undersea Fiber Optic Networks: Past, Present, and Future', *IEEE Journal on Selected Areas in Communications*, vol. 16, no. 7, September 1998, pp. 1220–25, available at <http://ieeexplore.ieee.org/iel4/ 49/15642/00725191.pdf?arnumber=725191>, accessed 4 March 2008; and '14 Tbps Over a Single Optical Fiber: Successful Demonstration of World's Largest Capacity', *NTT Press Release*, 29 September 2006, available at <http://www.ntt.co.jp/news/news06e/0609/060929a.html>, accessed 4 March 2008.

[39] Stephen Cass, 'Listening In', *IEEE Spectrum Special Report on Intelligence and Technology*, vol. 40, no. 4, April 2003, pp. 32–37, available at <http://www.estig.ipbeja.pt/ ~lmgt/st/other/Listening%20In.pdf>, accessed 4 March 2008; and 'NSA Tapping Underwater Fiber Optics', available at <http://slashdot.org/articles/01/05/23/2142216.shtml>, accessed 4 March 2008.

[40] John R. Freer, *Computer Communications and Networks*, UCL Press, University College London, London, 2nd edition, 1996, p. 305.

[41] Cass, 'Listening In', pp. 33–37.

[42] Department of Defence, *Force 2020*, p. 19.

[43] Gary Waters, 'Australia's Approach to Network Centric Warfare', in Waters and Ball, *Transforming the Australian Defence Force (ADF) for Information Superiority*, p. 14.

[44] Director General Capability and Plans, *NCW Roadmap 2007*, p. 5.

[45] Gary Waters, 'Australia's Approach to Network Centric Warfare', p. 14.

[46] See Philip Flood, *Report of the Inquiry into Australia's Intelligence Agencies*, Canberra, July 2004, available at <http://www.pmc.gov.au/publications/ intelligence_inquiry/index.htm>, accessed 4 March 2008.

[47] Vince Crawley and Amy Svitak, 'Is Predator the Future of Warfare?', *Defense News*, 11–17 November 2002, p. 8.

[48] Craig Hoyle and Andrew Koch, 'Yemen Drone Strike: Just the Start?', *Jane's Defence Weekly*, 13 November 2002, p. 3.

[49] Desmond Ball, 'Information Operations and Information Superiority', in Waters and Ball, *Transforming the Australian Defence Force (ADF) for Information Superiority*, p. 61.

[50] Gail Kaufman, 'New Eyes, New Rules', *Defense News*, 2–8 December 2002, pp. 1–2.

[51] Andrew Koch, 'New Powers for Info Operations Chiefs', *Jane's Defence Weekly*, 17 September 2003, p. 6.

[52] Robert Wall, 'Focus on Iraq Shapes Electronic, Info Warfare', *Aviation Week & Space Technology*, 4 November 2002, p. 34.

[53] Andrew Koch, 'Information Warfare Tools Rolled Out in Iraq', *Jane's Defence Weekly*, 6 August 2003, p. 7.

[54] David A. Fulghum, 'Infowar to Invade Air Defense Networks', *Aviation Week & Space Technology*, 4 November 2002, p. 30.

[55] Crawley and Svitak, 'Is Predator the Future of Warfare?', p. 8.

[56] Director General Capability and Plans, *NCW Roadmap 2007*, p. 21.

[57] Henry S. Kenyon, 'Networking Moves Into the High Frontier', *SIGNAL*, April 2004, pp. 59–62.

[58] 'Projects: JP 2008 Phase 3F—ADF SATCOM Capability Terrestrial Upgrade', 9 March 2007.

[59] Department of Defence, *Executive Summary. Draft Environmental Impact Statement (EIS): Defence Headquarters Australian Theatre*, Department of Defence, Canberra, September 2003, p. ES-8.

[60] Clay Wilson, *Information Operations and Cyberwar: Capabilities and Related Policy Issues*, Congressional Research Service, Library of Congress, Washington, DC, 14 September 2006, p. 8, available at <http://www.fas.org/irp/crs/RL31787.pdf>, accessed 4 March 2008.

[61] Alex Spillius, 'America Prepares for Cyber War with China', *Telegraph* (London), 15 June 2007, available at <http://www.telegraph.co.uk/news/ main.jhtml?xml=/news/2007/06/15/wcyber115.xml>, accessed 4 March 2008.

[62] Charles Bickers, 'Cyberwar: Combat on the Web', *Far Eastern Economic Review*, 16 August 2001, p. 30.

[63] Ivo Dawnay, 'Beijing Launches Computer Virus War on the West', *Age* (Melbourne), 16 June 1997, p. 8.

[64] Jason Sherman, 'Report: China Developing Force to Tackle Information Warfare', *Defense News*, 27 November 2000, pp. 1 and 19.

[65] Christopher Bodeen, 'Mainland Asks Taiwan to Stop Interference', *Washington Times*, 26 September 2002; and Doug Nairne, 'State Hackers Spying On Us, Say Chinese Dissidents', *South China Morning Post*, 18 September 2002, available at <http://www.infosyssec.com/securitynews/0209/6536.html>, accessed 4 March 2008.

[66] See, for example, 'Outrage in Berlin Over Chinese Cyber Attacks', 31 August 2007, available at <http://www.weeklystandard.com/weblogs/TWSFP/ 2007/08/outrage_in_berlin_over_chinese.asp>, accessed 4 March 2008.

[67] Yang Kuo-wen, Lin Ching-chuan and Rich Chang, 'Bureau Warns on Tainted Discs', *Taipei Times*, 11 November 2007, p. 2, available at <http://www.taipeitimes.com/ News/taiwan/archives/2007/11/11/2003387202>, accessed 4 March 2008.

[68] I-Ling Tseng, *Chinese Information Warfare (IW): Theory Versus Practice in Military Exercises (1996–2005)*, MA Sub-thesis, Graduate Studies in Strategy and Defence, Strategic and Defence Studies Centre, The Australian National University, Canberra, March 2005.

[69] 'Chinese Cyber Espionage "Routine" in Australia', *Canberra Times*, 11 February 2008, p. 5.

[70] 'MND Sets Up Information Warfare Committee', *ADJ News Roundup*, August 1999, p. 14.

[71] Francis Markus, 'Taiwan's Computer Virus Arsenal', *BBC News*, 10 January 2000, available at <http://news.bbc.co.uk/1/hi/world/asia-pacific/597087.stm>, accessed 4 March 2008; and Wendell Minnick, 'Taiwan Upgrades Cyber Warfare', *Jane's Defence Weekly*, 20 December 2000, p. 12.

[72] 'Taiwan to Conduct Cyber Warfare Drills', *Jane's Defence Weekly*, 16 August 2000, p. 10; Minnick, 'Taiwan Upgrades Cyber Warfare', p. 12; and Damon Bristow, 'Asia: Grasping Information Warfare?', *Jane's Intelligence Review, December 2000*, p. 34.

[73] Minnick, 'Taiwan Upgrades Cyber Warfare', p. 12.; and Darren Lake, 'Taiwan Sets Up IW Command', *Jane's Defence Weekly*, 10 January 2001, p. 17.

[74] Ministry of National Defense, *Republic of China, 2002 National Defense Report*, Ministry of National Defense, Taipei, July 2002. See also 'Taiwan Prepares for Cyber Warfare', *CNN.Com*, 29 July 2002; and 'Taiwan Report Finds Cyberthreat From China', *International Herald Tribune*, 30 July 2002.

[75] Chester Dawson, 'Cyber Attack', *Far Eastern Economic Review*, 10 February 2000, p. 21.

[76] Dawson, 'Cyber Attack'; and 'Japan/Crime: Cyber-terror Task Force Established', *Bangkok Post*, 27 January 2000, p. 6.

[77] 'Tokyo's Claim to Tok-do Escalates Korea-Japan Cyber War', *Korea Times*, 14 May 2000.

[78] Elaine Lies, 'Doomsday Cult Casts Shadow Over Japan', *Canberra Times*, 20 March 2000, p. 7.

[79] On January 2007, the Japan Defense Agency was upgraded to a Cabinet-level ministry, and is now known as the Japanese Ministry of Defense.

[80] Japan Defense Agency, *Defense of Japan 2000*, Japan Defense Agency, Tokyo, 2000, chapter 3, section 3(ii), and chapter 4, section 5(3). See also Damon Bristow, 'Asia: Grasping Information Warfare?', pp. 34–35.

[81] Juliet Hindell, 'Japan Wages "Cyber War" Against Hackers', 24 October 2000, *Internet Security News*, available at <http://www.landfield.com/isn/mail-archive/2000/Oct/0116.html>, accessed 4 March 2008.

[82] Bristow, 'Asia: Grasping Information Warfare?', p. 35.

[83] 'North Korea Ready to Launch Cyber War: Report', Computer Crime Research Center, 4 October 2004, available at <http://www.crime-research.org/news/04.10.2004/ North_Korea_ready_to_launch_cyber_war/>, accessed 4 March 2008.

[84] John Larkin, 'Preparing for Cyberwar', *Far Eastern Economic Review*, 25 October 2001, p. 64.

[85] Kevin Coleman, 'Inside DPRK's Unit 121', DefenseTech.org. 24 December 2007, available at <http://www.defensetech.org/archives/003920.html>, accessed 4 March 2008. See also 'North Korea

Operating Computer-hacking Unit', *Korea Herald*, 28 May 2004, available at <http://www.asiamedia.ucla.edu/article-eastasia.asp?parentid+11559>, accessed 4 March 2008.

[86] 'North Korea's Information Technology Advances and Asymmetric Warfare', *WMD Insights*, April 2006, available at <http://www.wmdinsights.org/ I4/EA1_NorthKoreaInfoTech.htm>, accessed 4 March 2008.

[87] Bristow, 'Asia: Grasping Information Warfare?', p. 36.

[88] Tim Huxley, *Defending the Lion City: The Armed Forces of Singapore*, Allen & Unwin, Sydney, 2000, p. 91.

[89] Bristow, 'Asia: Grasping Information Warfare?', p. 36.

Bibliography

Accenture, *The Accenture Security Practice: Security and the High-Performance Business*, 2003, available at <http://whitepapers.silicon.com/0,39024759, 60086441p,00.htm>, accessed 3 March 2008.

Ackerman, Robert K., 'Intelligence Center Mines Open Sources', *Signal*, March 2006, available at <http://www.afcea.org/signal/articles/templates/ SIGNAL_Article_Template.asp?articleid=1102&zoneid=31>, accessed 4 March 2008.

Alberts, David S., John J. Garstka, Richard E. Hayes, David A. Signori, *Understanding Information Age Warfare*, CCRP Publication Series, Washington, DC, August 2001, available at <http://www.dodccrp.org/ files/Alberts_UIAW.pdf>, accessed 4 March 2008.

Arquilla John, and David Ronfeldt, *The Advent of Netwar*, RAND Corporation, Santa Monica, CA, 1996.

Ashley, Bradley K. 'The United States is Vulnerable to Cyberterrorism', *SIGNAL*, March 2004, p. 61.

Australian Government, *Cybercrime Act: An Act to amend the law relating to computer offences, and other purposes*, No. 161, Canberra, 2001, available at <http://scaleplus.law.gov.au/html/comact/11/6458/pdf/ 161of2001.pdf>, accessed 11 March 2008.

Australian Government, Department of Defence, *A Concept for Enabling Information Superiority and Support*, Department of Defence, Canberra, August 2004.

———, *Australia's National Information Infrastructure: Threats and Vulnerabilities*, Defence Signals Directorate, February 1997.

———, *Australia's National Security: A Defence Update 2007*, Department of Defence, July 2007, available at <http://www.defence.gov.au/ans/2007/ pdf/Defence_update.pdf>, accessed 25 February 2008.

———, Director General Capability and Plans, *NCW Roadmap 2007*, Defence Publishing Service, Canberra, February 2007, available at <http://www.defence.gov.au/capability/ncwi/docs/2007NCW_Roadmap.pdf>, accessed 4 March 2008.

———, 'DSTO and ADI Forge New Links in Network Centric Warfare', Defence Science and Technology Organisation, 2 September 2004, available at <http://www.dsto.defence.gov.au/news/3283/>, accessed 4 March 2008.

———, *Enabling Future Warfighting: Network Centric Warfare*, ADDP-D.3.1, Australian Defence Headquarters, Canberra, February 2004, available at <http://www.defence.gov.au/strategy/fwc/documents/NCW_Concept.pdf>, accessed 4 March 2008.

———, *Executive Summary. Draft Environmental Impact Statement (EIS): Defence Headquarters Australian Theatre*, Department of Defence, Canberra, September 2003, p. ES-8.

———, *Explaining NCW*, Department of Defence, Canberra, 21 February 2006, available at <http://www.defence.gov.au/capability/NCWI/docs/Explaining_NCW-21feb06.pdf>, accessed 25 February 2008.

———, *Force 2020*, Department of Defence, Canberra, June 2002, available at <http://www.defence.gov.au/publications/f2020.pdf>, accessed 25 February 2008.

———, *Future Warfighting Concept*, Australian Defence Doctrine Publication (ADDP)-D.02, Department of Defence, Canberra, 2003, available at <http://www.defence.gov.au/publications/fwc.pdf>, accessed 25 February 2008.

———, 'Infosec', Defence Signals Directorate, available at <http://www.dsd.gov.au/ infosec/>, accessed 4 March 2008.

———, *Information Operations*, ADDP 3-13, Australian Defence Headquarters, 2006.

———, *Joint Operations for the 21st Century*, Department of Defence, Canberra, May 2007, available at <http://www.defence.gov.au/publications/FJOC.pdf>, accessed 25 February 2008.

———, *NCW Roadmap*, Department of Defence, Canberra, October 2005, updated in Director General Capability and Plans, *NCW Roadmap 2007*, Defence Publishing Service, Canberra, March 2007, available at <http://www.defence.gov.au/capability/ncwi/docs/2007NCW_Roadmap.pdf>, accessed 25 February 2008.

———, 'Network-Centric Warfare', Defence Science and Technology Organisation, available at <http://www.dsto.defence.gov.au/research/4051/page/4387/>, accessed 4 March 2008.

———, *The Australian Approach to Warfare*, Department of Defence, Canberra, June 2002, available at <http://www.defence.gov.au/publications/taatw.pdf>, accessed 4 March 2008.

Australian Government, Department of the Attorney-General, 'Budget 2001-2002 Fact Sheet. Protecting the National Information Infrastructure: Part of the Government's E-security Initiative'.

————, joint news release by the Minister for Communications, Information Technology and the Arts, and the Minister for Defence, 'Security in the Electronic Environment', 27 September 2001.

————, news release by Attorney-General Philip Ruddock MP, *Protecting Australia's Critical Infrastructure*, Parliament House, 11 May 2004, available at <http://ag.gov.au/www/agd/rwpattach.nsf/VAP/ (CFD7369FCAE9B8F32F341DBE097801FF)~MR03CriticalInfrastructure 07May04.doc/$file/MR03CriticalInfrastructure07May04.doc>, accessed 3 March 2008.

————, *Protecting Australia's National Information Infrastructure*, December 1998, available at <http://law.gov.au/publications/niirpt.ptl>.

Australian Government, Department of the Prime Minister and Cabinet, *Protecting Australia Against Terrorism 2006: Australia's National Counter-Terrorism Policy and Arrangements*, Department of the Prime Minister and Cabinet, Canberra, 2006, p. 60, available at <http://cipp.gmu.edu/archive/ Australia_ProtectAUTerrorism_2006.pdf>, accessed 3 March 2008.

Australian Homeland Security Research Centre, *National Security Briefing Notes—Advancing domestic and national security practice: 2007 E-Security National Agenda*, July 2007, available at <http://www.homelandsecurity. org.au/files/2007_e-security_agenda.pdf>, accessed 3 March 2008.

Ball, Desmond, *Australia's Secret Space Programs*, Canberra Papers on Strategy and Defence no. 43, Strategic and Defence Studies Centre, The Australian National University, Canberra, 1988, chapter 3.

————, *Signals Intelligence in the Post-Cold War Era: Developments in the Asia-Pacific Region*, Institute of Southeast Asian Studies, Singapore, 1993.

————, 'Silent Witness: Australian Intelligence and East Timor', in Richard Tanter, Desmond Ball and Gerry van Klinken, *Masters of Terror: Indonesia's Military and Violence in East Timor*, Rowman & Littlefield, New York, 2006, pp. 177–201.

Bamford, James, *Body of Secrets: How America's NSA and Britain's GCHQ Eavesdrop on the World*, Century, London, 2001.

Bean, Hamilton, 'The DNI's Open Source Center: An Organizational Communication Perspective', *International Journal of Intelligence and Counterintelligence*, vol. 20, no. 2, Summer 2007, pp. 240–57.

'Behind the Firewall—The Insider Threat', 15 April 2003, ARTICLE ID: 2122. See <http://enterprisesecurity.symantec.com/article.cfm? articleid=2122&PID= 14615847&EID=389>.

Bennett, Ralph, *Behind the Battle: Intelligence in the War with Germany 1939-1945*, Pimlico, London, 1999.

Best, Jr. Richard A. and Alfred Cumming, *Open Source Intelligence (OSINT): Issues for Congress*, CRS Report for Congress, Congressional Research Service, 5 December 2007, available at <http://www.fas.org/sgp/crs/intel/RL34270.pdf>, accessed 4 March 2008.

Bickers, Charles, 'Cyberwar: Combat on the Web', *Far Eastern Economic Review*, 16 August 2001, p. 30.

Blakely, Rhys, Jonathan Richards, James Rossiter, and Richard Beeston, 'MI5 Alert on China's Cyberspace Spy Threat', *TimesOnline*, 1 December 2007, available at <http://business.timesonline.co.uk/tol/business/industry_sectors/technology/article2980250.ece>, accessed 20 December 2007.

Blau, John, 'German Gov't PCs Hacked, China Offers to Investigate: China Offers to Help Track Down the Chinese Hackers Who Broke into German Computers', *PC World*, 27 August 2007, available at <http://www.washingtonpost.com/wp-dyn/content/article/2007/08/27/AR2007082700595.html>, accessed 4 March 2008.

Blaxland, John, 'On Operations in East Timor—The experiences of the Intelligence Officer, 3rd Brigade', *Australian Army Journal*, 2000.

Bodeen, Christopher, 'Mainland Asks Taiwan to Stop Interference', *Washington Times*, 26 September 2002.

Bristow, Damon, 'Asia: Grasping Information Warfare?', *Jane's Intelligence Review, December 2000*, p. 34–35.

Carnegie Mellon Software Engineering Institute, 'Cert/CC Statistics 1998-2005', undated.

Cass, Stephen, 'Listening In', *IEEE Spectrum Special Report on Intelligence and Technology*, vol. 40, no. 4, April 2003, pp. 32–37, available at <http://www.estig.ipbeja.pt/~lmgt/st/other/Listening%20In.pdf>, accessed 4 March 2008.

Cavelty, Myriam Dunn, 'Critical information infrastructure: vulnerabilities, threats and responses', in *Disarmament Forum* (Three), 2007, pp. 15–22, available at <http://se2.isn.ch/serviceengine/FileContent?serviceID=CRN&fileid=20009CBA-C36C-C7AC-D7C0-5E43B2974BC5&lng=en>, accessed 4 March 2008.

Cebrowski, Ret. Admiral Arthur, speech to Network Centric Warfare 2003 Conference, January 2003, available at <http://www.oft.osd.mil>, accessed 25 February 2008.

CERT, 'CERT Advisory CA-2001-14 Cisco IOS HTTP Server Authentication Vulnerability', 28 June 2001, available at <http://www.cert.org/advisories/CA-2001-14.html>, accessed 4 March 2008.

'Chinese Cyber Espionage "Routine" in Australia', *Canberra Times*, 11 February 2008, p. 5.

Chulov, Martin, 'A Win Against Terror', *Australian*, 7 October 2007, p. 17, available at <http://www.theaustralian.news.com.au/story/0,20867,20536940-5001561,00.html>, accessed 4 March 2008.

Cilluffo Frank J., and J. Paul Nicholas, 'Cyberstrategy 2.0', *Journal of International Security Affairs*, No. 10, Spring 2006, available at <http://www.securityaffairs.org/issues/2006/10/cilluffo_nicholas.php>, accessed 3 March 2008.

CISCO, 'Multiple Vulnerabilities in Cisco Secure Access Control Server', 7 January 2007, available at <http://www.securiteam.com/securitynews/5DP0420KAG.html>, accessed 4 March 2008.

———, 'Optus Charts Future with Cisco Service Oriented Network at Macquarie Park Campus', 19 October 2006, available at <http://newsroom.cisco.com/dlls/global/asiapac/news/2006/pr_10-19.html>, accessed 4 March 2008.

———, 'CISCO Security Advisories', available at <http://www.cisco.com/en/US/products/products_security_advisories_listing.html>, accessed 4 March 2008.

Clark, Drew, 'Computer security officials discount chances of "digital Pearl Harbor"', *National Journal's Technology Daily*, 3 June 2003, available at <http://www.govexec.com/dailyfed/0603/060303td2.htm>, accessed 3 March 2008.

Coleman, Kevin, 'Inside DPRK's Unit 121', DefenseTech.org. 24 December 2007, available at <http://www.defensetech.org/archives/003920.html>, accessed 4 March 2008.

Computer Fraud and Abuse Act, 1986 (18 USC 1030) (amended 1994, 1996, and 2001).

Cornford, Philip, and Rob O'Neill, 'Bali Nine Phone Cards Cracked', *Age*, 4 May 2005.

Cosgrove, Peter, 'Innovation, People, Partnerships: Continuous Modernisation in the ADF'; speech to the Network Centric Warfare Conference on 20 May 2003, available at <http://www.defence.gov.au/cdf/speeches/past/speech20030520.htm>, accessed 25 February 2008.

Council of Europe Treaty Office, *Convention on Cybercrime*, CETS No. 185, opened for signature in Budapest, Hungary on 23 November 2001, available at

<http://conventions.coe.int/Treaty/Commun/QueVoulezVous.asp?
NT=185&CM=8&DF=16/04/04&CL=ENG>, accessed 11 March 2008.

Counterfeit Access Device and Computer Fraud and Abuse Act, 1984 (Public Law 99-474).

Crawley, Vince and Amy Svitak, 'Is Predator the Future of Warfare?', *Defense News*, 11–17 November 2002, p. 8.

Crompton, Malcolm, 'Proof of ID Required? Getting Identity Management Right', keynote address to the Australian IT Security Forum, Sydney, 30 March 2004, available at <http://www.privacy.gov.au/news/speeches/sp1_04p.pdf>, accessed 3 March 2008.

Datz, Todd, 'Out of Control', *CSO*, vol. 2, no. 1, 2005, p. 28.

Davis, Mark, 'Canberra, CEOs extend forum on Terrorism', *Australian Financial Review*, 24 June 2004, p. 3.

Dawnay, Ivo, 'Beijing Launches Computer Virus War on the West', *Age* (Melbourne), 16 June 1997, p. 8.

Dawson, Chester, 'Cyber Attack', *Far Eastern Economic Review*, 10 February 2000, p. 21.

de Nysschen, Heinrich, 'Homeland Security', *Image & Data Manager*, May/June 2005, p. 36.

Denning, Dorothy E., 'Activism, Hacktivism, and Cyberterrorism: The Internet as a Tool for Influencing Foreign Policy', in J. Arquilla and D. Ronfeldt (eds), *Networks and Netwars: The Future of Terror, Crime, and Militancy*, RAND Corporation, Santa Monica, CA, 2001, pp. 239–88, available at <http://www.rand.org/pubs/monograph_reports/MR1382/MR1382.ch8.pdf>, accessed 4 March 2008.

———, *Obstacles and Options for Cyber Arms Controls*, paper presented at Arms Control in Cyberspace Conference, Heinrich Böll Foundation, Berlin, 29–30 June 2001, available at <http://www.cs.georgetown.edu/~denning/infosec/ berlin.doc>, accessed 4 March 2008.

Dudgeon, Ian, 'Intelligence Support to the Development and Implementation of Foreign Policies and Strategies', *Security Challenges*, vol. 2, no. 2, July 2006, pp. 61–80, available at <http://securitychallenges.org.au/SC%20Vol%202%20No%202/vol%202%20no%202%20Dudgeon.pdf>, accessed 26 February 2008.

Fidler, Stephen, 'Steep Rise in Hacking Attacks from China', *Financial Times*, 5 December 2007, available at <http://www.ft.com/cms/s/0/c93e3ba2-a361-11dc-b229-0000779fd2ac.html>, accessed 4 March 2008.

'Fighting the worms of mass destruction', *Economist*, 27 November 2003, available at <http://www.economist.com/science/displayStory.cfm?story_id= 2246018>, accessed 3 March 2008.

Filippidis, Arthur, Tan Doan and Brad Tobin, 'Net Warrior—DSTO Battlelab Interoperability', Simulation Industry Association of Australia, June 2007, available at <http://www.siaa.asn.au/simtect/ 2007/Abstracts/70.html>, accessed 4 March 2008.

Flood, Philip, *Report of the Inquiry into Australia's Intelligence Agencies*, Canberra, July 2004, available at <http://www.pmc.gov.au/publications/ intelligence_inquiry/index.htm>, accessed 4 March 2008.

'14 Tbps Over a Single Optical Fiber: Successful Demonstration of World's Largest Capacity', *NTT Press Release*, 29 September 2006, available at <http://www.ntt.co.jp/news/news06e/0609/060929a.html>, accessed 4 March 2008.

Freer, John R., *Computer Communications and Networks*, UCL Press, University College London, London, 2nd edition, 1996.

Fulghum, David A., 'Infowar to Invade Air Defense Networks', *Aviation Week & Space Technology*, 4 November 2002, p. 30.

Gale, Stephen, *Protecting Critical Infrastructure*, Foreign Policy Research Institute, November 2007, available at <http://www.fpri.org/enotes/200711.gale. infrastructure.html>, accessed 4 March 2008.

Goldman, Emily O., 'New Threats, New Identities, and New Ways of War: The Sources of Change in National Security Doctrine', *Journal of Strategic Studies*, vol. 24, no. 2, 2001, pp. 43–76.

Gonsalves, Antone, 'Gartner: Dependence On Internet Boosts Risks of Cyberwar', *InformationWeek*, 15 January 2004, available at <http://www.information week.com/story/showArticle.jhtml?articleID=17301666>, accessed 3 March 2008.

Government of Canada, Office of Critical Infrastructure Protection and Emergency Preparedness, Threat Analysis No. TA03-001, 12 March 2003, available at <http://www.ocipep-bpiepc.gc.ca/opsprods/other/TA03-001 e.pdf>, accessed 4 March 2008.

Grabosky, Peter, *Electronic Crime*, Prentice Hall, Upper Saddle River, NJ, 2007.

Greene, Kate, 'Calling Cryptographers', *MIT Technology Review*, 16 February 2006, available at <http://www.technologyreview.com/Infotech/ 16347/?a=f>, accessed 4 March 2008.

Greenfield, Heather, 'Industry Officials Sketch Priorities for DHS Cyber Czar', *National Journal's Technology Daily*, 2 October 2006, available at

<http://www.govexec.com/dailyfed/1006/100206tdpm1.htm>, accessed 3 March 2008.

Griffith, Samuel B., *Sun Tzu: The Art of War*, Oxford University Press, London, Oxford and New York, 1963, p. 77.

'Half the world has a mobile phone', *ITU News*, January–February 2008.

Hindell, Juliet, 'Japan Wages "Cyber War" Against Hackers', *Internet Security News*, 24 October 2000, available at <http://www.landfield.com/ isn/mail-archive/2000/Oct/0116.html>, accessed 4 March 2008.

Homer-Dixon, Thomas, 'The Rise of Complex Terrorism', *Foreign Policy*, Issue No. 128, January//February 2002, pp. 52–62.

Hoyle, Craig and Andrew Koch, 'Yemen Drone Strike: Just the Start?', *Jane's Defence Weekly*, 13 November 2002, p. 3.

Huxley, Tim, *Defending the Lion City: The Armed Forces of Singapore*, Allen & Unwin, Sydney, 2000.

Tseng, I-Ling, *Chinese Information Warfare (IW): Theory Versus Practice in Military Exercises (1996–2005)*, MA Sub-thesis, Graduate Studies in Strategy and Defence, Strategic and Defence Studies Centre, The Australian National University, Canberra, March 2005.

'Increased Telephone Interception Capacity', in Australian Federal Police, *National Illicit Drug Strategy Initiatives, November 1997—April 2001* (Second edition), p. 13, available at <http://www.afp.gov.au/ __data/assets/pdf_file/6634/ nids.pdf>, accessed 4 March 2008.

International Institute for Strategic Studies, *International Institute for Strategic Studies (IISS) Strategic Survey 2003/4*, Oxford University Press, Oxford, May 2004.

International Organization for Standardization, *Information technology—Security techniques—Code of practice for information security management*, ISO/IEC, Second edition, Geneva, Switzerland, 16 June 2005.

'Japan/Crime: Cyber-terror Task Force Established', *Bangkok Post*, 27 January 2000, p. 6.

Japan Defense Agency, *Defense of Japan 2000*, Japan Defense Agency, Tokyo, 2000, chapter 3, section 3(ii), and chapter 4, section 5(3).

Javelin Strategy & Research, *2007 Identity Fraud Survey Report: Identity Fraud is Dropping, Continued Vigilance Necessary*, Pleasanton, CA, February 2007.

Jenkins, Chris, 'Internet Terrorism Fears as Virus Hits', *Australian*, 28 January 2004, p. 3.

Kaufman, Gail, 'New Eyes, New Rules', *Defense News*, 2–8 December 2002, pp. 1–2.

Kellner, Mark A., 'China a "Latent Threat, Potential Enemy": Expert', *DefenseNews Weekly*, 4 December 2006, available at <http://www.defensenews.com/ story.php?F=2389588&C=america>, accessed 4 March 2008.

Kenyon, Henry S., 'Networking Moves Into the High Frontier', *SIGNAL*, April 2004, pp. 59–62.

Kerfoot, Frank W., and William C. Marra, 'Undersea Fiber Optic Networks: Past, Present, and Future', *IEEE Journal on Selected Areas in Communications*, vol. 16, no. 7, September 1998, pp. 1220–25, available at <http://ieeexplore.ieee.org/iel4/49/15642/00725191.pdf?arnumber=725191>, accessed 4 March 2008.

Knights, Michael, 'Options for Electronic Attack in the Iraq Scenario', *Jane's Intelligence Review*, December 2002, pp. 52–53.

Koch, Andrew, 'Information Warfare Tools Rolled Out in Iraq', *Jane's Defence Weekly*, 6 August 2003, p. 7.

———, 'New Powers for Info Operations Chiefs', *Jane's Defence Weekly*, 17 September 2003, p. 6.

Krasner, Stephen D. (ed.), *International Regimes*, Cornell University Press, Ithaca, New York, 1983, p. 2.

Lake, Darren, 'Taiwan Sets Up IW Command', *Jane's Defence Weekly*, 10 January 2001, p. 17.

Landler, Mark, and John Markoff, 'In Estonia, What May Be the First Cyberwar', *International Herald Tribune*, 28 May 2007, available at <http://www.iht.com/bin/print.php?id=5901141>, accessed 4 March 2008.

Larkin, John, 'Preparing for Cyberwar', *Far Eastern Economic Review*, 25 October 2001, p. 64.

Lentz, Robert F., testimony before the House Armed Services Committee on Terrorism, Unconventional Threats and Capabilities, hearing on 'Cyber Terrorism: The New Asymmetric Threat', 24 July 2003, available at <http://www.iwar.org.uk/cip/resources/status-of-dod-ia/03-07-24lentz.htm>, accessed 3 March 2008.

Libicki, Martin C., *What is Information Warfare?*, Center for Advanced Concepts and Technology, National Defense University, Washington, DC, 1995.

Lies, Elaine, 'Doomsday Cult Casts Shadow Over Japan', *Canberra Times*, 20 March 2000, p. 7.

Lonsdale, David J., *The Nature of War in the Information Age: Clausewitzian Future*, Frank Cass, London and New York, 2004.

Lowenthal, Mark M., *Intelligence, From Secrets to Policy*, Second Edition, CQ Press, Washington, DC, 2003.

Mahony, Trevor W., 'A Hybrid Civilian/Military Payload to Support Battlefield Communications', *Journal of Battlefield Technology*, vol. 1, no. 1, March 1998, pp. 29–32.

Makarenko, Tamara, 'The Crime-Terror Continuum: Tracing the Interplay Between Transnational Organised Crime and Terrorism', *Global Crime*, vol. 6, no.1, February 2004, pp. 129–45, available at <http://www.silkroadstudies.org/new/docs/publications/ Makarenko_GlobalCrime.pdf>, accessed 26 February 2008.

Markoff, John, 'China Link Suspected in Lab Hacking', *New York Times*, 9 December 2007, p. A-03, available at <http://www.nytimes.com/ 2007/12/09/us/nationalspecial3/09hack.html>, accessed 4 March 2008.

Markus, Francis, 'Taiwan's Computer Virus Arsenal', *BBC News*, 10 January 2000, available at <http://news.bbc.co.uk/1/hi/world/ asia-pacific/597087.stm>, accessed 4 March 2008.

McCarthy, John A., 'Introduction: From Protection to Resilience: Injecting 'Moxie' into the Infrastructure Security Continuum', in *Critical Thinking: Moving from Infrastructure Protection to Infrastructure Resilience*, CIP Program Discussion Paper Series, George Mason University, Washington, DC, 2007, pp. 2–3, available at <http://cipp.gmu.edu/archive/CIPP_ Resilience_Series_ Monograph.pdf>, accessed 4 March 2008.

McGee, J.V., L. Prusak, and P.J. Pyburn, *Managing Information Strategically: Increase your Company's Competitiveness and Efficiency by Using Information as a Strategic Tool*, John Wiley & Sons, New York, 1993.

McGlinchey, David, 'Agencies, Congress urged to upgrade computer security planning', GovExec.com, Washington DC, 17 March 2004, available at <http://www.govexec.com/dailyfed/0304/031704d1.htm>, accessed 3 March 2008.

McKenna, Tim, Terry Moon, Richard Davis and Leoni Warne, 'Science and Technology for Australian Network-Centric Warfare: Function, Form and Fit', *ADF Journal*, no. 17, pp. 62–75.

McLachlan, Kevin, 'Flaw Found in Cisco Secure Access Control Server', 26 June 2006, available at <http://www.crn.com/it-channel/189601708>, accessed 4 March 2008.

Merritt, Ira W., 'Proliferation and Significance of Radio Frequency Weapons Technology', Statement before the Joint Economic Committee, US

Congress, Washington, DC, 25 February 1998, available at
<http://www.house.gov/ jec/hearings/radio/merritt.htm>, accessed 4
March 2008.

Milliken, Robert, 'Canberra Acts to Keep an Eye on its Spies', *Independent*
(London), 2 June 1995, available at <http://findarticles.com/p/articles/
mi_qn4158/is_19950602/ai_n13986087>, accessed 4 March 2008.

Ministry of National Defense, *Republic of China, 2002 National Defense Report*,
Ministry of National Defense, Taipei, July 2002.

Minnick, Wendell, 'Computer Attacks from China leave many questions', *Defense
News*, 13 August 2007, available at <http://www.taiwanmilitary.org/
phpBB2/viewtopic.php?p=38438&sid=8f527c809bde63b7c174fd9b3fbdb7dd>,
accessed 4 March 2008.

————, 'Taiwan Upgrades Cyber Warfare', *Jane's Defence Weekly*, 20 December
2000, p. 12.

'MND Sets Up Information Warfare Committee', *ADJ News Roundup*, August
1999, p. 14.

Näf, Michael, 'Ubiquitous Insecurity? How to "Hack" IT Systems', *Information
& Security: An International Journal*, no. 7, 2001, pp. 104–18, available
at <http://se1.isn.ch/serviceengine/FileContent?serviceID=
PublishingHouse&fileid=9F1EA165-76C6-BF34-7522-
6D4EA03FB0F5&lng=en>, accessed 4 March 2008.

Nairne, Doug, 'State Hackers Spying On Us, Say Chinese Dissidents', *South China
Morning Post*, 18 September 2002, available at
<http://www.infosyssec.com/ securitynews/0209/6536.html>, accessed
4 March 2008.

National Commission on Terrorist Attacks upon the United States, *The 9/11
Commission Report*, W.W. Norton & Company, Inc., New York, 2004,
available at <http://www.9-11commission.gov/report/911Report.pdf>,
accessed 3 March 2008.

'North Korea Operating Computer-hacking Unit', *Korea Herald*, 28 May 2004,
available at <http://www.asiamedia.ucla.edu/article-eastasia.asp?parentid
+11559>, accessed 4 March 2008.

'North Korea Ready to Launch Cyber War: Report', Computer Crime Research
Center, 4 October 2004, available at <http://www.crime-research.org/
news/04.10.2004/North_Korea_ready_to_launch_cyber_war/>, accessed
4 March 2008.

'North Korea's Information Technology Advances and Asymmetric Warfare',
WMD Insights, April 2006, available at

<http://www.wmdinsights.org/I4/ EA1_NorthKoreaInfoTech.htm>, accessed 4 March 2008.

'Now France Comes Under Attack from PRC Hackers', *Agence France Presse*, 9 September 2007, available at <http://www.taipeitimes.com/News/front/ archives/2007/09/09/2003377917>, accessed 4 March 2008.

'NSA Tapping Underwater Fiber Optics', available at <http://slashdot.org/ articles/01/05/23/2142216.shtml>, accessed 4 March 2008.

'Optus Positions for National Satellite Success', December 2001, available at <http://www.optus.net.au/portal/site/aboutoptus/menuitem. 813c6f701cee5a14f0419f108c8ac7a0/?vgnextoid= a7ab8336054f4010VgnVCM1000009fa87c0aRCRD&vgnextchannel= b93cfaf924954010VgnVCM10000029a67c0aRCRD&vgnextfmt=default>, accessed 4 March 2008.

'Outrage in Berlin Over Chinese Cyber Attacks', 31 August 2007, available at <http://www.weeklystandard.com/weblogs/TWSFP/2007/08/ outrage_in_berlin_over_chinese.asp>, accessed 4 March 2008.

Palowitch, Andrew, 'Cyber Warfare: Viable Component to the National Cyber Security Initiative?' speech delivered at Georgetown University, Washington, DC, 27 November 2007.

Perry, William G., 'Enhanced data mining information assurance by using ISO 17799', *Information: Assurance and Security, Data Mining, Intrusion Detection, Information Assurance and Data Networks Security*, Defense & Security Symposium, The International Society for Optical Engineering, 17 April 2006.

———, *Information Warfare: An Emerging and Preferred Tool of the People's Republic of China*, Occasional Papers Series, no. 28, October 2007, The Center for Security Policy, Washington, DC, available at <http://www.centerforsecuritypolicy.org/modules/newsmanager/ center%20publication%20pdfs/perry%20china%20iw.pdf>, accessed 4 March 2008.

———, 'The Science of Protecting the Nation's Critical Infrastructure', *Voices of Discovery*, Elon University, NC, 7 March 2007.

Potts (ed.), David, *The Big Issue: Command and Combat in the Information Age*, Strategic and Combat Studies Institute Occasional Paper no. 45, CCRP Publication Series, February 2003, pp. 244–45, available at <http://www.dodccrp.org/files/Potts_Big_Issue.pdf>, accessed 3 March 2008.

'Privacy exposed', *Sydney Morning Herald*, 19 February 2004, available at <http://smh.com.au/articles/2004/02/18/1077072702295.html>, accessed 3 March 2008.

'Red storm rising: DoD's efforts to stave off nation-state cyberattacks begin with China', *Government Computer News*, 21 August 2006, available at <http://www.gcn.com/print/25_25/41716-1.html>, accessed 4 March 2008.

Reed, Donald J., 'Why Strategy Matters in the War on Terror', *Homeland Security Affairs*, vol. II, no. 3, October 2006, p. 5, available at <http://www.hsaj.org/ pages/volume2/issue3/pdfs/2.3.10.pdf>, accessed 3 March 2008.

Richelson, Jeffrey, 'Desperately Seeking Signals', *Bulletin of the Atomic Scientists*, vol. 56, no. 2, March/April 2000, pp. 47–51.

Ronfeldt, David F., and John Arquilla, *Networks and Netwars*, RAND Corporation, Santa Monica, CA, January 2002.

Sabo, John T., *Addressing a Critical Aspect of Homeland Security: Managing Security and Privacy in Information Sharing Systems*, Computer Associates White Paper, January 2004, available at <http://www.ehcca.com/ presentations/privacyfutures1/4_01_2.pdf>, accessed 4 April 2008.

Sands, Amy , 'Integrating Open Sources into Transnational Threat Assessments', in Jennifer E. Sims and Burton Gerber, *Transforming U.S. Intelligence*, Georgetown University Press, Washington, DC, 2005, p. 64.

Saxton, Jim, opening statement before the House Armed Services Committee on Terrorism, Unconventional Threats and Capabilities; hearing on 'Cyber Terrorism: The New Asymmetric Threat', 24 July 2003, available at <http://www.iwar.org.uk/cip/resources/status-of-dod-ia/03-07-24saxton.htm>, accessed 3 March 2008.

Schmidtchen, David, *The Rise of the Strategic Private: Technology, Control and Change in a Network Enabled Military*, The General Sir Brudenell White Series, Land Warfare Studies Centre, Canberra, 2006.

Schriner, David, 'The Design and Fabrication of a Damaging RF Weapon by "Back Yard" Methods', Statement before the Joint Economic Committee, US Congress, Washington, DC, 25 February 1998, available at <http://www.house.gov/jec/ hearings/02-25-8h.htm>, accessed 4 March 2008.

Shen, Dan, Genshe Chen, Jose B. Cruz, Jr., Erik Blasch, and Martin Kruger, *Game Theoretic Solutions to Cyber Attack and Network Defense Problems*, paper given to 12th ICCRTS Conference, entitled 'Adapting C2 to the 21st

Century', 2007, available at <http://www.dodccrp.org/events/ 12th_ICCRTS/CD/html/papers/062.pdf>, accessed 4 March 2008.

Sherman, Jason, 'Report: China Developing Force to Tackle Information Warfare', *Defense News*, 27 November 2000, pp. 1 and 19.

Singh, Ajay, 'Time: The New Dimension in War', *Joint Force Quarterly*, no. 10, Winter 1995–96, pp. 56–61, available at <http://www.dtic.mil/ doctrine/jel/jfq_pubs/1510.pdf>, accessed 4 March 2008.

Sipress, Alan, 'An Indonesian's Prison Memoir Takes Holy War Into Cyberspace', *Washington Post*, 14 December 2004, p. A19, available at <http://www.washingtonpost.com/wp-dyn/articles/A62095-2004Dec13.html>, accessed 4 March 2008.

Smith, Edward A., *Complexity, Networking & Effects-Based Approaches to Operations*, Command and Control Research Program (CCRP), Department of Defense, July 2006, available at <http://www.dodccrp.org/files/ Smith_Complexity.pdf>, accessed 26 February 2008.

Spafford, Eugene. H., testimony before the House Armed Services Committee on Terrorism, Unconventional Threats and Capabilities; hearing on 'Cyber Terrorism: The New Asymmetric Threat', 24 July 2003, available at <http://www.iwar.org.uk/cip/resources/status-of-dod-ia/03-07-24 spafford.pdf>, accessed 3 March 2008.

Spillius, Alex, 'America Prepares for Cyber War with China', *Telegraph* (London), 15 June 2007, available at <http://www.telegraph.co.uk/news/ main.jhtml?xml=/news/2007/06/15/wcyber115.xml>, accessed 4 March 2008.

Stewart, Cameron, 'Telstra Operation Helped Track Down Bali Bombers', *Australian*, 7 October 2006, p. 8, on-line version entitled 'Telstra Secretly helped Hunt Bali Bombers' at <http://www.news.com.au/story/0,23599, 20537904-2,00.html>, accessed 4 March 2008.

Stiftung, Heinrich Böll, *Perspectives for Peace Policy in the Age of Computer Network Attacks*, Conference Proceedings, 2001, available at <http://www.boell.de/downloads/medien/DokuNr20.pdf>, accessed 4 March 2008.

Szafranski, Colonel R., 'A Theory of Information Warfare: Preparing for 2020', *Airpower Journal*, vol. 9, no. 1, Spring 1995, available at <http://www.iwar.org.uk/iwar/resources/airchronicles/szfran.htm>, accessed 4 March 2008.

'Taiwan Prepares for Cyber Warfare', *CNN.Com*, 29 July 2002.

'Taiwan Report Finds Cyberthreat From China', *International Herald Tribune*, 30 July 2002.

'Taiwan to Conduct Cyber Warfare Drills', *Jane's Defence Weekly*, 16 August 2000, p. 10.

'Telecommunications Interception Law Dispute Shows Law Needs Overhaul', *Electronic Frontiers Australia*, 31 March 2004, available at <http://www.efa.org.au/Publish/PR040331.html>, accessed 4 March 2008.

Tenpas, Ronald J., Statement of Associate Deputy Attorney General before the Subcommittee on Terrorism, Technology and Homeland Security the Committee on the Judiciary, 21 March 2007, available at <http://judiciary.senate.gov/testimony. cfm?id=2582&wit_id=6194>, accessed 4 March 2008.

Thompson, Clive, 'The Virus Underground', *New York Times*, 8 February 2004, available at <http://engineering.dartmouth.edu/courses/engs004/virusarticle.html>, accessed 26 February 2008.

'Timeline of Notable Computer Viruses and Worms', Wikipedia, available at <http://en.wikipedia.org/wiki/Timeline_of_notable_computer_viruses_and_worms>, accessed 4 March 2008.

Titheridge, Alan, Gary Waters, and Ross Babbage, *Firepower to Win: Australian Defence Force Joint Fires in 2020*, Kokoda Paper no. 5, The Kokoda Foundation, Canberra, October 2007.

Tkacik, John J. Jr., *Trojan Dragons: China's International Cyber Warriors*, WebMemo no. 1735, The Heritage Foundation, 12 December 2007, available at <http://www.heritage.org/Research/AsiaandthePacific/upload/wm_1735.pdf>, accessed 4 March 2008.

Toffler, Alvin and Heidi, *War and Anti-War: Survival at the Dawn of the 21st Century*, Little Brown, London, 1994.

'Tokyo's Claim to Tok-do Escalates Korea-Japan Cyber War', *Korea Times*, 14 May 2000.

United States Department of Defense, *Department of Defense Dictionary of Military and Associated Terms*, Joint Publication 1-02, 17 October 2007, available at <http://www.dtic.mil/doctrine/jel/doddict/data/g/02329.html>, accessed 28 February 2008.

———, *Information Operations*, Joint Publication 3-13, 13 February 2006, available at <http://www.dtic.mil/doctrine/jel/new_pubs/jp3_13.pdf>, accessed 4 March 2008.

———, *Joint Doctrine for Information Operations*, Joint Publication 3-13,9 October 1998, available at <http://www.iwar.org.uk/iwar/resources/us/jp3_13.pdf>, accessed 4 March 2008.

————, *Report on Network Centric Warfare*, 2001, available at <http://www.defenselink.mil/nii/NCW/ncw_sense.pdf>, accessed 25 February 2008.

United States Department of Homeland Security, Remarks by Assistant Secretary Gregory Garcia at the RSA Conference on IT and Communications Security, San Francisco, CA, 8 February 2007, available at <http://www.dhs.gov/xnews/speeches/sp_1171386545551.shtm>, accessed 4 March 2008.

United States Government Accountability Office (GAO), *CYBERCRIME: Public and Private Entities Face Challenges in Addressing Cyber Threats*, GAO-07-705, Report to Congressional Requesters, Washington, DC, June 2007, available at <http://www.gao.gov/new.items/d07705.pdf>, accessed 4 March 2008.

————, *Technology Assessment: Cybersecurity for Critical Infrastructure Protection*, GAO-04-321, Washington, DC, 28 May 2004, available at <http://www.gao.gov/ new.items/d04321.pdf>, accessed 4 March 2008.

Uniting and Strengthening America by Providing Appropriate Tools Required to Intercept and Obstruct Terrorism (USA PATRIOT) Act, 2001 (Public Law 107-56).

US White House, *Critical Infrastructure Identification, Prioritization, and Protection*, Homeland Security Presidential Directive No. 7, White House, 17 December 2003, available at <http://www.whitehouse.gov/news/ releases/2003/ 12/20031217-5.html>, accessed 3 March 2008.

————, *The National Strategy to Secure Cyberspace*, White House, Office of the Press Secretary, February 2003, available at <http://www.whitehouse. gov/pcipb/cyberspace_strategy.pdf>, accessed 4 March 2008.

'Use It But Don't Lose It', *Aviation Week & Space Technology*, 9 September 2002, p. 29.

van Loon, J. 'Virtual Risks in an Age of Cybernetic Reproduction', in B. Adam, U. Beck and J. van Loon (eds), *The Risk Society and Beyond: Critical Issues for Social Theory*, Sage, London, 2000.

Wall, Robert, 'Focus on Iraq Shapes Electronic, Info Warfare', *Aviation Week & Space Technology*, 4 November 2002, p. 34.

Waltz, Edward, *Information Warfare: Principles and Operations*, Artech House Publications, Boston and London, 1998.

Waters, Gary and Desmond Ball, *Transforming the Australian Defence Force (ADF) for Information Superiority*, Canberra Papers on Strategy and Defence no. 159, Strategic and Defence Studies Centre, The Australian National University, Canberra, 2005.

Wilson, Clay, *Information Operations and Cyberwar: Capabilities and Related Policy Issues*, Congressional Research Service, Library of Congress, Washington, DC, 14 September 2006, p. 8, available at <http://www.fas.org/irp/crs/RL31787.pdf>, accessed 4 March 2008.

———, *Network Centric Warfare: Background and Oversight Issues for Congress*, Congressional Research Service (CRS) Report for Congress, 2 June 2004, available at <http://www.fas.org/man/crs/RL32411.pdf>, accessed 25 February 2008.

Yang Kuo-wen, Lin Ching-chuan and Rich Chang, 'Bureau Warns on Tainted Discs', *Taipei Times*, 11 November 2007, p. 2, available at <http://www.taipeitimes.com/News/taiwan/archives/2007/11/11/2003387202>, accessed 4 March 2008.

Yoshihara, Toshi, *Chinese Information Warfare: A Phantom Menace or Emerging Threat?*, Strategic Studies Institute, U.S. Army War College, Carlisle, PA, November 2001, available at <http://www.strategicstudiesinstitute.army.mil/ pdffiles/PUB62.pdf>, accessed 4 March 2008.

Websites

Armed Forces Communications and Electronics Association: <http://www.afcea.org/>

Australian Computer Emergency Response Team: <http://www.auscert.org.au/>

Australian Federal Police: <http://www.afp.gov.au/home.html>

Australian Government Attorney General's Department: <http://ag.gov.au/>

Australian Government Department of Defence: <http://www.defence.gov.au/>

Australian Government Department of Defence, Defence Signals Directorate: <http://www.dsd.gov.au/>

Australian Government Department of Defence, Defence Science and Technology Organisation: <http://www.dsto.defence.gov.au/>

Australian Government Department of the Prime Minister and Cabinet: <http://www.pmc.gov.au/>

Australian Government Initiative, Stay Smart Online: <http://www.staysmartonline.gov.au/>

Australian Government Office of the Privacy Commissioner: <http://www.privacy.gov.au/>

Australian High Tech Crime Centre: <http://www.ahtcc.gov.au>

Australian Homeland Security Research Centre: <http://www.homelandsecurity.org.au/>

Center for Security Policy: <http://www.centerforsecuritypolicy.org/>

CERT Coordination Center: <http://www.cert.org/certcc.html>

Cisco Systems, Inc: <http://www.cisco.com/>

Computer Crime Research Center: <http://www.crime-research.org/>

Council of Europe Treaty Office: <http://conventions.coe.int/>

Electronic Frontiers Australia: <http://www.efa.org.au/>

Federation of American Scientists: <http://www.fas.org/>

Foreign Policy Research Institute: <http://www.fpri.org/>

International Organization for Standardization:
     <http://www.iso.org/iso/home.htm>

Military.com, DefenseTech.org: <http://www.defensetech.org/>

Rand Corporation: <http://www.rand.org/>

Simulation Industry Association of Australia: <http://www.siaa.asn.au/>

Strategic Studies Institute of the U.S. Army War College:
     <http://www.strategicstudiesinstitute.army.mil/>

The Heritage Foundation: <http://www.heritage.org/>

United States Department of Defense: <http://www.defenselink.mil/>

United States Department of Defense, Chief Information Officer, Assistant
     Secretary of Defense (Networks & Information Integration):
     <http://www.defenselink.mil/cio-nii/>

United States Department of Defense, Command and Control Research Program:
     <http://www.dodccrp.org/>

United States Department of Defense, Defense Technical Information Center:
     <http://www.dtic.mil/>

United States Department of Defense, Office of Force Transformation:
     <http://www.oft.osd.mil>

United States Department of Homeland Security:
     <http://www.dhs.gov/index.shtm>

United States Government Accountability Office: <http://www.gao.gov/>

United States House of Representatives: <http://www.house.gov/>

United States House of Representatives Joint Economic Committee:
     <http://www.house.gov/jec/>

Index